333
SCIENCE
TRICKS &
EXPERIMENTS

This volume is dedicated to a man who worked with the project since he was 5 years old. He is my son, Robert J. Brown, Jr. He is a scientist in industry, yet still enjoys our elementary and junior high school tricks and experiments. He originated many of the experiments and improved many others.

It is also dedicated to my daughter, Betty Brown Yow, who began performing science tricks and experiments for me before she was 8 years old. Their assistance and encouragement have been invaluable through the years. Family cooperation could not have been possible without the love and understanding of my wife, Mary T. Brown.

No. 1825
$15.95

Brown, Robert J.

333

SCIENCE
TRICKS &
EXPERIMENTS

ROBERT J. BROWN

TAB **TAB BOOKS Inc.**
BLUE RIDGE SUMMIT, PA 17214

FIRST EDITION

FIRST PRINTING

Copyright © 1984 by TAB BOOKS Inc.

Printed in the United States of America

Library of Congress Cataloging in Publication Data

Brown, Robert J., 1907-
    333 science tricks and experiments.

    Includes index.
     √1. Science—Experiments. √2. Scientific recreations.
I. Title.   II. Title: Three hundred thirty-three science
tricks and experiments.
√Q164.B8387   1984   507'.8   84-8875
ISBN 0-8306-0825-7
ISBN 0-8306-1825-2 (pbk.)

Line drawings by Frank W. Bolle.
Technical assistance by Arthur Wood.

Contents

Introduction

This book contains some "Science for You" experiments that have run in newspapers through the L.A. Times Syndicate. More experiments can be found in 333 More Science Tricks and Experiments (TAB book No. 1835). The book brings interest and delight to young and old alike who have a natural and investigative curiosity about how and why things happen.

This volume can be especially rewarding to teachers and students of science through junior high, fathers and mothers whose kids always need something challenging to do, prospective Science Fair contestants searching for good project ideas, and, not the least important, young people who seek desperately to understand the world around them. Many, if given the chance and preparation, will be the scientists and engineers of tomorrow—who just may save mankind from itself.

Many of the experiments are thought to have originated with the author. Many others are improvements and simplifications of existing experiments. Some are corrections of errors found in the explanations of experiments in other sources. The author has sought to minimize the possibility of error; he has three erudite and meticulous consultants who check everything. Comments and suggestions are always welcome as the author presents more and more experiments through his column.

Chapter 1

Inertia & Momentum

THE ELASTIC COINS

NEEDED: Five nickels and a copper penny.

EXPERIMENT 1: Flip one coin against another, as in the upper drawing, and the elasticity of the metal will be demonstrated by the bouncing of the coins. Flip the penny hard against the pile of nickels, and one nickel will fly out from the bottom of the stack as shown below.

EXPERIMENT 2: Try a quarter. If the quarter strikes head-on and at the right velocity, the bottom nickel flies out and the rest of the nickels come to rest on the top of the quarter. If the

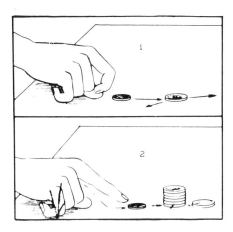

The Elastic Coins

1

quarter strikes at a grazing angle, the nickel flies out at the bottom and the quarter goes off at an angle.

REASON: The coins are not pure soft metal, but alloys that increase hardness and the elasticity. Therefore, they bounce. The size of the coins is important. The "copper penny" referred to here is not an English penny, but the standard American cent piece which is part copper. The penny must be thinner than a nickel, so as not to hit the second nickel in the stack. A dime will work if it is flipped hard enough.

In the second experiment, the inertia of the stack of nickels keeps it practically still while the bottom coin is flipped out from under the pile.

Simple experiments such as these can involve mathematics if the experimenter wants to go that far into them. Other words that may be used in their interpretation can include impact, impulse, and conservation of momentum.

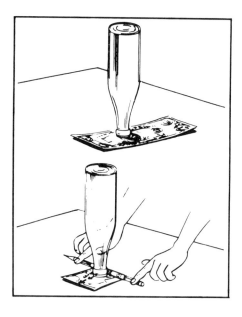

A Dollar Bill Puzzle

A DOLLAR BILL PUZZLE

NEEDED: A dollar bill, an empty bottle, and a pencil.

EXPERIMENT 1: This is one of the old-time puzzlers: take the bill from under the bottle without touching the bottle or tipping it over. It's easy if the bill is rolled around the pencil and the bottle pushed aside in that way.

EXPERIMENT 2: Try jerking the dollar bill quickly while the bottle rests on it. The bottle will probably not tip over; its inertia tends to hold it in place. Do not raise the bill as you jerk.

EXPERIMENT 3: Pull the dollar bill slowly. The bottle will probably move with it. Pull a little more quickly, and the bottle will probably tip over.

REASON: A light push on the upper part of the bottle will create enough torque to turn the bottle over. It is not very stable in this inverted position because of its high center of gravity.

A similar push at the lower part of the bottle causes a much smaller torque but causes enough force to slide the bill from under the bottle.

Soft drink bottles are good for these experiments. They must be dry.

CENTER OF GRAVITY

NEEDED: A long pencil and a short one.

EXPERIMENT 1: Stand the pencils upright, and release both at the same time. The short one falls to the table much faster than the long one.

EXPERIMENT 2: Notice that when the pencil has fallen and is horizontal it moves in the direction of the fall. The mass center has some horizontal velocity because the pencil was forced to fall on the arc of a circle.

EXPERIMENT 3: Watch a small child fall from a standing position. He almost never gets hurt. The fall is so short that the speed gained is not great; he does not hit the floor very hard. If he were much taller, or fell from a table, gravity would have more time to accelerate him (speed him up), and he might hit the floor hard enough to hurt himself.

REASON: Gravity is the force which acts on both pencils to bring them to a horizontal position. The center of gravity of the longer pencil is much higher above the table than that of the shorter; therefore, in the longer pencil, more mass must fall through a longer distance with a longer arc. This takes more time.

BALANCE A HATCHET

NEEDED: A hammer, hatchet, axe or garden hoe.

EXPERIMENT 1: Try to balance the hatchet with the handle up. It is difficult if not impossible. Try balancing it with the heavy end up. It is easy.

EXPERIMENT 2: Try this with a broom or a long pole. It is much easier to balance a long object for the same reason. Circus performers take advantage of this principle when they balance one another on chairs on long poles.

EXPERIMENT 3: Note that the circus wire walker balances himself with a long pole that bows down at the ends. Try this (near the ground!) and see if it is easy. If the pole bows down far enough it will be self-balancing.

REASON: The hatchet will try to fall in either case, but if the heavy end is up, there is more inertia to hold it in position so you have more time to move the hand to counteract the falling movement.

Chapter 2

Sound & Other Vibrations

HOW FAR THE STORM?

NEEDED: A distant thunderstorm (or other source of sound) and a stop watch.

EXPERIMENT: Start the count as the lightning is seen, and stop when the sound of the thunder is heard. The length of time taken for the sound to reach the ear tells the distance of the lightning flash.

REASON: Sound travels in air about 1100 feet per second. If it takes five seconds for a sound to reach the ear, the distance is about 5500 feet, a little more than a mile.

The light from the flash travels so fast (more than 186,000 miles per second) that its speed need not be considered in the calculations.

A good way to estimate the distance if there is no stop watch is to count "thousand one, thousand two, thousand three," etc., so that each count would be a second, representing about one-fifth of a mile.

A TRICKY HUMMER

NEEDED: A cardboard tube and a tall jar of water.

EXPERIMENT: Place the tube in the water, and hold it with one hand while holding the jar with the other. Whistle or hum over the open end of the tube, keeping a constant note. Raise and lower the jar of water so that the water in the tube goes up and down

A Tricky Hummer

slowly, thus making the air column in the tube longer and shorter.

When the correct point is reached, the sound is much louder.

REASON: When the point of resonance of the air column is reached, the air vibrates at the frequency of the whistle. At this point, each vibration of the air column reinforces the vibration of the whistle sound, making it louder. This is a state of resonance, and the vibrations are called "sympathetic" vibrations.

If resonance is not obtained from the first whistle or hum tone, try a different pitch. Also try a different position of the mouth at the end of the tube.

KNIFE AND FORK CHIMES

NEEDED: A knife, a fork, a spoon, some string.

EXPERIMENT 1: Tie the silver pieces together so that they do not touch each other. Hold the ends of the string to the ears, and

Knife and Fork Chimes

as the head is moved and the silver pieces clang together, there is the sound of beautiful chimes.

REASON: The sounds heard are very much like the sounds of ordinary clanking of the silver, except that each sound lasts longer since the silver is free to vibrate. The string conducts the sounds to the ears, making them louder and more mellow.

The mellowness is accounted for by the softness of the string, which filters out some of the harshness. The vibrations in the individual pieces are at regular frequencies, and these give musical tones. Irregular vibrations would make noise.

EXPERIMENT 2. Replace the string with rubber bands tied together. Hold the rubber to the ears as the string was held. There will likely be no sound at all, certainly not chimes. This is because rubber is not elastic in the scientific sense.

REASON: Vibrations from the silverware are fed into the string, and travel up the string to the ears. The string is elastic enough to transmit the vibrations with a little loss. In the rubber, the sound energy introduced in the stretching in each vibration is not completely given back in the relaxation, but some of it is turned into heat.

The sound wave gets weaker as it goes up the rubber, and soon dies out completely.

The common definition of "elastic" is "stretchable-but-finally-coming-back." So, in common usage, rubber and things woven

Irregular Vibrations

of rubber are elastic. In the scientific sense, glass and hard steel are very elastic, while rubber is not.

IRREGULAR VIBRATIONS

NEEDED: A drinking glass, a pencil, a string.

EXPERIMENT: Tie the string tightly around the glass and loosely around the pencil. Hold the pencil and let the glass hang down. As the pencil is turned, a peculiar noise comes from the glass.

REASON: The string does not slide easily on the pencil when the pencil is turned, but turns with the pencil for a short distance, then slips back. The jerky movement makes vibrations in the string, and the string transfers them to the glass. The glass sets up vibrations in the air within and around the glass, and these reach the ear.

The Open and Shut Pipe

THE OPEN AND SHUT PIPE

NEEDED: A large-diameter soda straw and a sharp razor blade.

EXPERIMENT: Cut the straw at about the middle. Flatten one end of one piece of the straw. Arrange the pieces as shown and blow through the straw. A musical tone will be heard. Remove your finger from the end of the lower straw and blow again. The new musical tone will be an octave higher.

Try different lengths of the lower straw.

REASON: This is a condition of resonance (look up the word) of an air column in a pipe as in organ pipes or other wind instrument.

The air column in the closed tube vibrates a maximum amount at the open end (antinode) and none at the closed end (node). In the open tube there is maximum vibration at each end (antinode) with a node between them. The wave length in the closed tube is two times that in the open tube. Therefore, the frequency in the closed tube is one half that in the open tube.

A Musical Rubber Band

A MUSICAL RUBBER BAND

NEEDED: A tin can, two nails, a rubber band

EXPERIMENT: Drive the nails half way into the can, as shown, and stretch the rubber band between them. Pluck the band with the finger and a sound will be heard. Wrap the band around the nails a few times, pluck it, and the pitch heard when the band is plucked will be higher.

REASON: As the band is plucked, it is set into vibration. The greater the tension on the band, the faster the vibration, and the more rapid the vibration, the higher the pitch.

The can serves to hold the nails and to vibrate with the nails. The air in the can vibrates, too, and may make the sound louder.

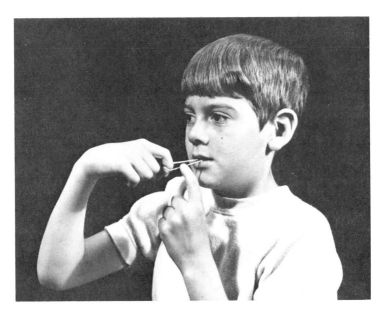
Hear Through the Teeth

HEAR THROUGH THE TEETH

NEEDED: A rubber band.

EXPERIMENT: Hold the band stretched between the teeth and a finger, pluck it, and a musical sound is heard. Have someone else hold the end that was between the teeth, pluck it again, and the sound is not as loud.

REASON: The teeth and jaws act somewhat as a sounding board, vibrating with the vibrations in the rubber band. The sound waves are conducted directly to the ears through the bones of the skull. When the band does not touch the teeth only the sound waves produced in the air by the vibrating band reach the ears through the air, as in normal hearing. Also, the soft flesh of the fingers reduces the sound when hands are used instead of teeth.

A MYSTERY SOUND

NEEDED: A glass, a fork, a tabletop without a cloth.

EXPERIMENT: Have someone sit at the edge of the table with an ear to the drinking glass. Strike the fork. Touch the handle of the fork to the tabletop. Suddenly the sound of the vibrating fork seems to be coming from the glass.

REASON: If the fork is held above the table, a slight sound of the vibrating fork may be heard directly through the air. But if the

11

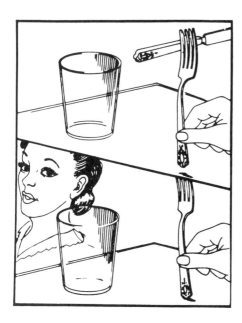

A Mystery in Sound

fork is touched to the tabletop, the table vibrates much as the sounding board of a musical instrument. The vibrations are carried from the fork to the table, to the glass, and into the air surrounding the glass, where the sound seems much louder.

Holding the ear over the glass makes the sound louder than if the ear is held an equal distance above the table because of resonance vibrations of the air column in the glass itself. Different-sized glasses may be tried.

WHAT CAN CARRY SOUND?

NEEDED: A tabletop, a bucket of water, a comb.

EXPERIMENT 1: Put your ear on the table, and have someone tap the table at the other end. The sound is carried to the ear through the tabletop.

EXPERIMENT 2: Put your ear to the bucket of water, and have someone rub his fingers over the teeth of the comb while it is under the water. The sound will be carried through the water to your ear.

REASON: Most sounds come to us as vibrations in the air. But they can come through many substances—not only water and wood, but through the bones of the head and our teeth as well. They travel least easily through soft, sound-deadening substances.

Musical Balloon

MUSICAL BALLOON

NEEDED: A rubber balloon.

EXPERIMENT: Blow up the balloon. Place it under the arm and pull the neck of the balloon as shown in drawing at upper left.

By varying the pressure and the stretch, musical (?) tones can be produced.

REASON: Air escaping through the stretched rubber neck of the balloon will not flow steadily, because the rubber expands and contracts, causing the air to come out in a series of puffs or waves. If these are irregular, they produce noise. If they come in regular intervals, they can make rather pleasant musical sounds.

SODA-STRAW MUSIC

NEEDED: Large-diameter straws, scissors.

EXPERIMENT: Press the straw end with the fingers. With a little practice, you can blow into the straw and produce a musical note. Clipping off the straw makes the pitch higher.

REASON: Practice is necessary, as the lips must be taught to vibrate properly. This is necessary in the playing of some of the musical wind instruments.

Soda Straw Music

The length of the straw controls the length of the air column in the straw. The pitch is varied by the length of the vibrating air column.

THE NOISY CAN

NEEDED: A tin can, a string, rosin, a pencil, a matchstick.

EXPERIMENT: Make a hole in the middle of the can, on the bottom, and tie the string on the matchstick inside the can. Have someone hold the can while you draw the rosin back and forth on the string.

The rosin will make an unpleasant sound. When the pencil is

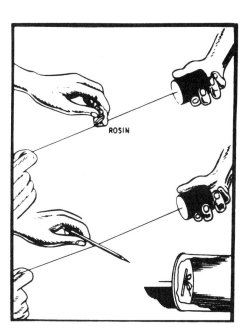

The Noisy Can

then drawn along the string, it, too, will make the unpleasant sound. Remember that a similar movement of the violin bow across the strings can make beautiful music.

REASON: The rosin sticks, releases, sticks, and releases time after time. This makes a jerky movement in the string, which is transmitted to the can as irregular vibrations. The rosin makes the pencil move at irregular intervals over the string.

The vibrations of string, pencil, and can are transmitted through the air to the ear, and are noise. In the tuned violin, the vibrations come at regular intervals, and so can be music!

THE BULLDOG

NEEDED: A piece of thin wood such as a ruler, a strong cord, a round stick.

EXPERIMENT: Tie the cord to a hole in the end of the ruler. Tie the other end of the cord to a notch in the round stick—loosely so it can turn on the stick, Whirl the ruler around overhead, and listen to the bulldog growl!

REASON: As the thin wood moves through the air it is caught by air currents and twirled. As it twirls it creates disturbances in the air that reach the ear as regalar but rather discordant sounds.

The tone is changed with the speed at which the wood is

twirled, and is different with different shaped pieces of wood. The operation of this little toy is complex, and its explanation could start many an argument in a college physics class.

Be careful the wood does not fly out of control and hit someone.

Chapter 3

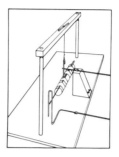

Projects to Build

A VIBRATION DETECTOR

NEEDED: A good flashlight cell, three dead cells, pencil sharpener, earphones, wires and connecting clips, wood, woodworking tools.

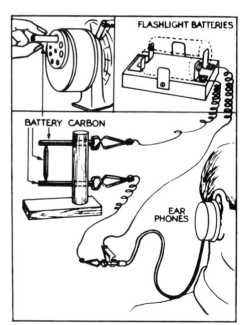

A Vibration Detector

EXPERIMENT: Take a carbon out of a dead cell, and sharpen it at both ends. Bore holes in the other two carbons from the dead cells, so they will hold the sharpened one when set into dowels. The sharpened carbon must rest loosely in the holes. Connect as shown. One flashlight cell will give enough power.

OBSERVATION: Set the instrument on the ground, and ground vibrations may be heard. If the earphones are in another room, speech may be heard coming over the device.

REASON: Air movement or vibration causes the carbon-to-carbon contact to be better or worse, varying the amount of current that may pass through it. This is the principle of the telephone microphone, which is called the "transmitter." Carbon granules are used in it. (Idea suggested by Russell Richner, Lake Worth, Florida.)

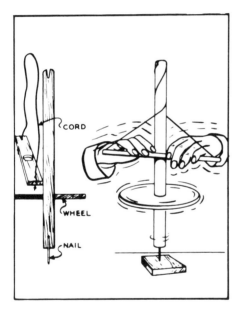

The Indian Drill

THE INDIAN DRILL

NEEDED: A broom handle, a nail, a string, pieces of wood, some tools.

EXPERIMENT: Assemble the pieces as shown in drawing at left so that the wooden handle will be stopped by the string just above the flywheel. Start the drill by winding some of the string as shown in drawing above. By moving the handle up and down, the drill will be made to spring rapidly back and forth. The nail will drill a hole in the wooden block.

REASON: The principle of this instrument is similar to that of the yo-yo, in that the pulling of the string starts the instrument in a rapid turning motion. Then its momentum will carry it on, while the string winds around the broomstick in the other direction.

This device has been used by many primitive tribes in many parts of the world to drill holes and start fires. Because it was used by some of the American Indians it has been called "The Indian Drill."

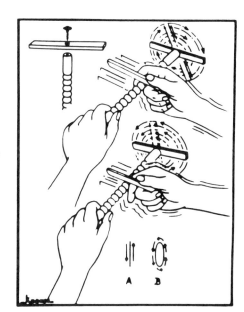

A Magic Propeller

A MAGIC PROPELLER

NEEDED: Wood, a screw, a knife.

EXPERIMENT: Carve a stick so that the notches on it are evenly spaced. They do not have to go all the way around the stick; the trick will work if they are cut on only two edges. Mount the propeller on the end with the screw, so that it is loose and turns freely.

Rub one stick with the other, while rubbing the finger against the notches. The propeller turns. Place the finger on the other side of the notched stick, and the propeller reverses and turns in the opposite direction.

REASON: If the notches are rubbed with the stick alone, the vibrations in the stick are likely to be straight back and forth, as shown in small diagram A. The propeller will not turn.

When the stick and fingers both rub the notches, the vibration takes a circular path, as in diagram B. The propeller turns because of this circular vibration. Practice makes perfect!

Make Your Own Alcohol Lamp

MAKE YOUR OWN ALCOHOL LAMP

NEEDED: A jar with a metal cover, a large nail, some cloth for a wick.

EXPERIMENT: Punch a hole in the lid with the big nail as shown. Drive the nail into wood from the upside-down lid, so that the projecting metal in the lamp will point upward. Put the wick in, put alcohol in the jar, and you have a lamp that will give a hot flame.

The metal lid will conduct heat away from the flame, so that the flame does not go below the lid into the alcohol. But any fire is dangerous, so be careful.

A HOMEMADE TELEPHONE

NEEDED: Two paper cups, string.

EXPERIMENT 1: Make holes in the bottoms of the cups, put string through the holes, and tie short sticks to the ends to hold the

A Homemade Telephone

strings in place. Stretch the string—you have an improved "tin can telephone."

This is an old experiment when tin cans are used. It will be found that paper cups work better, because the paper vibrates more easily than the metal of a can.

Use hard twisted string; soft cotton string will transmit less of the vibrations from one cup to the other.

EXPERIMENT 2: If fine wire is available, try it in place of the cord. If the wire is fine it should conduct the sound. If it is heavy its inertia will kill some of the vibrating.

MEASURE AIR MOISTURE

NEEDED: A board, a peg, some paper-backed foil, glue.

EXPERIMENT: Mount the peg in the board. Cut a strip out of the foil, glue the end of the strip to the peg, then wind the strip of foil around the peg.

REASON: The coil of foil will unwind somewhat, then will wind and unwind slightly more as the moisture content of the air varies. The paper in the strip absorbs moisture and expands. The foil does not. The paper will force the foil to bend slightly as it expands and contracts.

This crude instrument cannot, of course, be depended on as a hygrometer or an air moisture measuring device. But it illustrates the principle of one type of such device.

Measure Air
Moisture

MAKE YOUR OWN KALEIDOSCOPE

NEEDED: A 5 × 7 dime store mirror, a way to cut it, plastic wrap or tissue paper, rubber bands, various colored soda straws.

EXPERIMENT: Cut the mirror lengthwise into three pieces (they will be about 1 5/8 × 7) and cut tiny pieces of soda straw with scissors. Hold the three glass pieces together (as shown) with rubber bands, and put the paper or plastic over one end. Hold it on with a rubber band.

Hold the device up to the eye as shown, turn it slowly, and many beautiful patterns will be seen in the plastic pieces as they fall—never two patterns exactly the same.

REASON: The explanation is the multiple reflections of the colored bits in the three mirrors. Each mirror reflects them at least two times.

BUILD YOUR OWN SCALE

NEEDED: Wood, a can lid, a hacksaw blade, screws, string.

EXPERIMENT: Make the scale or balance as shown, a hacksaw blade as the spring (suggested by the Nuffield Junior Science Apparatus book).

Make Your Own
Kaleidoscope

To calibrate the author's model he used weights of one-half, one, two, and four ounces. It is suggested that metric weights be used if available.

Build Your Own
Scale

Build Your Own Scale

OBSERVATION: This scale is quite sensitive; its usefulness is determined mainly by the accuracy and exactness of the calibration.

An instrument such as this is often called a "balance." This is not technically correct. A balance compares masses hung on each end of a lever; no spring is used.

THE DIODE (1)

NEEDED: A diode, a lamp, socket, battery, all to match, a double-pole, double-throw switch, wire for connections.

EXPERIMENT: Connect the parts as shown in the diagram. When the switch is closed either way, the battery is connected to the lamp, but the lamp will burn one way and not the other.

REASON: A diode is a strange instrument that will let current flow in one direction through it, but not the other. Reversing the switch reverses the polarity of the battery, which means that it reverses the direction of the current flow.

Diodes, along with transistors, are the "solid state" devices that have revolutionized electronics in the past few years. They do tasks that formerly had to be performed with vacuum tubes.

Parts needed for this experiment may be bought from a radio supply store.

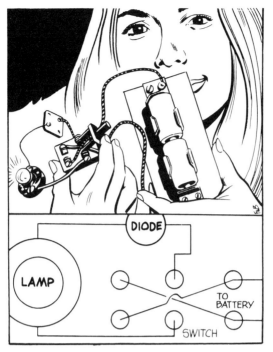

The Diode (1)

THE DIODE (2)

NEEDED: A diode, a lamp, a single pole, double-throw switch, a transformer, wire for connections.

EXPERIMENT: Connect the parts as shown, so that when the switch is closed in one position the current can flow directly to the lamp, but when closed in the other position the current must pass through the diode. The lamp is dim when the current has to flow through the diode.

REASON: A diode allows current to flow in one direction through it. Current coming from a transformer flows in one direction, then reverses and flows in the other direction, so about half the current is stopped by the diode and does not reach the lamp. The arrow points to the diode in the drawing.

Parts may be obtained from a radio and television store. Almost any diode will do for this, and they are not expensive. Some toy train transformers have rectifiers built in so that they give

25

The Diode (2)

current that flows one direction (dc) and these will not work in this experiment. Straight (ac) current is required.

A FLYING SAUCER DETECTOR

NEEDED: A strong bar magnet, copper wire, brass screws, wood, tape, thread.

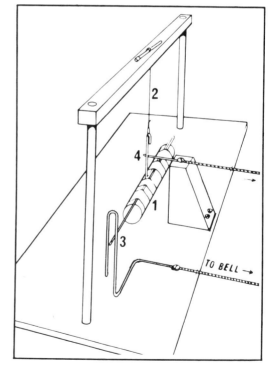

A Flying Saucer Detector

EXPERIMENT: Build the detector as shown. The magnet and some of the wire are put together by wrapping tape around them, as shown in No. 1. Part of the bare wire forms a loop or hook. A thin thread, 2, hooks into the wire loop to suspend the magnet so it can turn freely.

A bare wire touches the hook or loop at 4, making electrical contact. At 3 is a loop that is touched as the magnet and wire turn slightly, making another electrical contact.

Set the instrument where it cannot be moved and where it will not get dusty (cover it!). There must not be a draft or breeze on it. Move it around until the magnet, pointing north and south, comes to rest so the extended wire is between the wires of loop 3, not touching. Keep all iron away.

OBSERVATION: A variation in the earth's magnetic field, caused by a flying saucer or a geomagnetic storm, will turn the magnet slightly, making contact at 3 and ringing the bell.

This instrument does not guarantee the appearance of a flying saucer; it does not mean that flying saucers exist. But it will make a good construction project.

(Adapted from a design by John Oswald, North Hampton. N. H.).

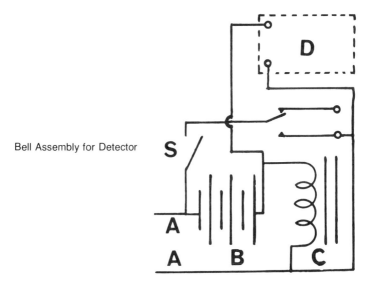

Bell Assembly for Detector

BELL ASSEMBLY FOR DETECTOR

NEEDED: Wires, bell, battery, switch, relay.
EXPERIMENT: Assemble the parts as shown. An elec-

tronics or electrical man will help, and this will be a good introduction to electrical circuit diagrams. The bell is shown at D, the relay at C, battery at B, and A shows the wires leading from the flying saucer detector previously mentioned.

An electronics store is the place to get the parts, and the man there will help select them so they match.

OBSERVATION: When the flying saucer detector magnet is disturbed, it turns slightly, making contact to short out the wires shown at A and A. The bell starts to ring.

At the same time the relay operates to close its contacts, providing a permanent short to keep the bell ringing. The bell is silenced by opening the switch S.

As stated, this flying saucer detector is not guaranteed to detect a flying saucer. But it is an interesting project for a boy or girl for the science fair or just for fun at home or at school.

The "Psychic" Motor

THE "PSYCHIC" MOTOR

NEEDED: Typing paper, a needle, a slender bottle.
EXPERIMENT: Construct the "motor" as shown, balancing

it with the needle point resting on the top of the bottle. Hold the hand close to it, and it will begin to turn mysteriously.

REASON: Martin Gardner, columnist for the *Scientific American,* presented this as a psychic motor. It was an April Fool joke, but fooled many people. He explains that the "motor" can be turned by air currents in the room, the breath, and convection currents caused by heat from the hand.

Gardner got the idea from a 1923 issue of *Science and Invention* magazine. The mystery motion is mysterious mainly because the motor is practically frictionless and can be run by imperceptible air currents.

A Potato Pop-Gun

POTATO POP-GUN

NEEDED: Potatoes, a metal or plastic tube with thin walls, a dowel, a knife.

EXPERIMENT: Cut potato slices a half to an inch thick. Press the tube down on them to cut out disks that will fit into the tube like corks. Place one in each end of the tube, push the dowel in *quickly,* and a potato "bullet" shoots out.

REASON: When the dowel is pushed in, the motion of the potato in the tube compresses the air between it and the potato disk at the other end of the tube. The compressed air forces the disk out, acting as a compressed spring.

The author, in building his pop-gun, used a piece of the thin overflow pipe found in toilet tanks. This came from a scrap box in a plumbing shop. For a dowel, he used the solid end of an old wooden window shade roller. The pipe had to be reamed to make the inside smooth at the ends where it was cut. The dowel should be nearly as large in diameter as the tube.

Many proprietors of plumbing shops are glad to help a boy or girl with a project in science.

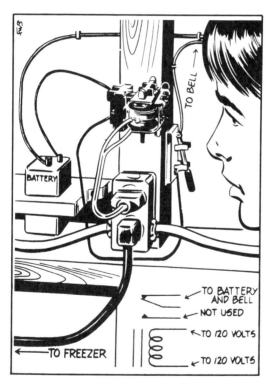

A Power-Off Alarm

A POWER-OFF ALARM

NEEDED: One relay, Porter & Brumfield GA11A, cord and plug, wire, bell, battery, switch.

EXPERIMENT: Connect the parts as shown. Keep the plug in. If the power goes off, the bell rings.

REASON: The electricity from the wall outlet keeps the relay energized. If the power fails the relay turns on the battery current to ring the bell.

If the GA11A relay is not found a substitute will do, but get an electronics man to help with the connections. Be sure the relay is in a metal box, since its bare 120-volt terminals are dangerous.

Such an alarm might prevent tardiness at school or might prevent spoilage of food in a freezer if the power is off long enough. The bell may be turned off by opening the switch, but it will stop by itself when the current is on again.

Chapter 4

Tricks

A PEPPER TRICK

NEEDED: A glass of water, pepper in a shaker, soap.

EXPERIMENT: Sprinkle pepper on the surface of the water, draw the finger across it as shown in upper drawing, and the pepper will separate where the finger has been.

Challenge a friend to do the same. The pepper will close in behind the finger as shown in lower drawing.

REASON: Before drawing your finger across the water surface, secretly rub your finger on a bar of soap. The soap weakens the

A Pepper Trick

surface film of the water. Without the soap, the surface tension or film draws the pepper back to cover the surface of the water more rapidly.

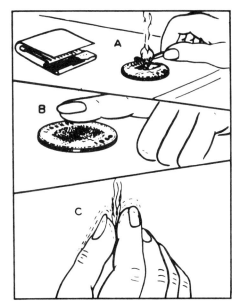

Smoke from the Fingertips

SMOKE FROM THE FINGERTIPS

NEEDED: A book of safety matches and a half dollar.

EXPERIMENT: Tear off a thin piece of the striking surface as shown at A, place it on the coin, and burn it with a match as shown.

Rub the finger on the darkened place on the coin, as in B. Then, when the fingers are rubbed together as in C, a mysterious smoke will rise from them.

REASON: The striking surface of a safety match book contains free red phosphorus. The heat of the hands and the heat of friction from the rubbing causes a very slight union of phosphorus with oxygen of the air to form a white vapor of phosphorous oxide.

A HOLE IN THE HAND

NEEDED: A cardboard mailing tube or a tube made of paper.

EXPERIMENT: Look through the tube at a distant object, placing the tube at the left eye. Bring the right hand up beside the tube. A hole will be seen in the hand.

REASON: The eyes see two images, but these are combined

in the brain so that we are not confused. In this case we are somewhat confused; the right eye sees the hand, and the left eye sees the distant object, and these are not compatible in the brain.

It is an optical illusion.

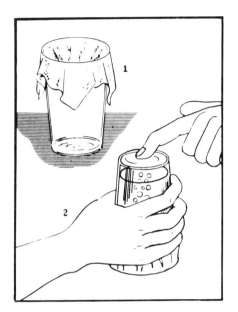

A Gypsy Mystery

A GYPSY MYSTERY

NEEDED: A glass of water and a cloth.

EXPERIMENT: Place the cloth (a cotton handkerchief will do) over the glass of water, turn it upside down, and the water will not run out. Have someone ask a question, and if the answer is "yes" the water will begin to "boil" when the finger is placed on the glass as shown in 2.

REASON: When the cloth is placed over the glass, be sure it is pushed downward, as shown in Fig. 1. Hold the cloth when the glass is inverted, and when the finger presses down on the glass, let the glass slip downward into the cloth if you want the answer to be "yes."

Of course, the water does not actually boil, but seems to as the cloth is pulled tighter, because air bubbles will come through the cloth and float to the surface of the water.

MOVING MATCHES

NEEDED: Five wooden matches, a drop of water.

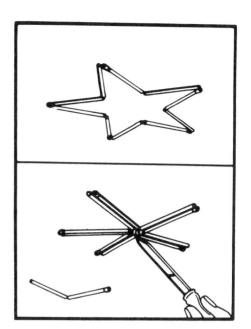

Moving Matches

EXPERIMENT: Break each match in the center and arrange them as shown, so that the broken places touch. Let a drop of water fall so that all the breaks are touched by the water. The matches should not be broken completely in two parts.

Slowly, the matches will begin to move until they have formed a star.

REASON: This is an example of capillary action. The water moves by capillary action into the dry wood, swelling the cells. The swelling tends to make the matches straighten out somewhat.

IT'S MAGIC!

NEEDED: A polyethylene bag, a sharp pencil.

EXPERIMENT: Blow up the bag like a balloon, and twist the opening to make it air tight. Push a sharp pencil through the bag, in one side and out the other, leaving ends protruding.

The balloon does not burst and the air does not leak out.

REASON: Polyethylene plastic will seal itself around the pencil because of its peculiar molecular structure.

The thick polyethylene bags suitable for this experiment are found in grocery stores. Raw carrots are sold in them, and sometimes other vegetables. They are also used for wrapping frozen foods.

CAUTION: Keep all plastic bags away from small children.

SECRET WRITING

NEEDED: A porcelain tabletop or sheet of glass, two sheets of paper, some water, a hard lead pencil with a smooth rounded point.

EXPERIMENT: Wet one sheet of paper, place it on the smooth surface, put the dry paper over the wet one, and write with firm pressure on the dry paper.

The writing can be seen on the wet paper, but will disappear when the paper is dry. It reappears when the paper is wet again.

REASON: The pressure of the pencil on the paper compresses the wet fibers so that they reflect light in a slightly different way when wet. The writing will be dim, like the watermark on some stationery.

A Paper Trick

A PAPER TRICK

NEEDED: A card, a rule, a pencil, a sharp knife or razor blade.

EXPERIMENT: Cut the card as shown in the drawing at upper left. The card can then be opened out, so that the hole in it is large enough for a person to step through.

A piece of strong paper the size of a postal card, 3 1/4 × 5 1/2 inches, or even much smaller, may be used, in place of the card.

JUMPING BEANS

NEEDED: A capsule from a drugstore and a shot or small ball bearing.

EXPERIMENT: Put the ball into the capsule, place it on the hand, and as the hand is moved, the capsule flips over and over in a lively manner.

REASON: The mass of the ball times the velocity it acquires in free motion when the capsule is tipped to slope somewhat downward is called its momentum. The weight of the light hollow capsule is not sufficient to stop the ball, so it flips over as the ball reaches the rounded end.

Real "jumping beans" are entirely different. They jump because of the movement of a tiny live animal inside.

HINDU MAGIC

NEEDED: A glass jar with a narrow opening, a table knife, enough uncooked rice to fill the jar.

EXPERIMENT: Fill the jar with rice, and plunge the knife into it a few times. Then announce that you will cause the jar to rise. Now plunge the knife deep into the rice, raise it slowly, and the entire jar will be lifted.

REASON: Strangely enough, there is no hidden secret, although the trick has been performed by Hindu fakirs and called magic. As the knife is stuck a dozen times into the rice, it packs the rice more and more tightly. When the blade is finally plunged deep into the rice, the grains have been packed tightly enough to press against the blade with enough frictional force to lift the jar.

FIREWORKS FROM A LEMON

NEEDED: A candle flame and a lemon.

EXPERIMENT 1: Squeeze the lemon peel near the flame as shown in the drawing, and small displays of "fireworks" may be seen shooting from the flame.

REASON: As the lemon peel is bent, some of the oil and water in it squirt out into the flame. Some of the oil burns as it passes through the flame, and some of the water vaporizes and sputters.

EXPERIMENT 2: Sprinkle flour on the candle flame. Tiny sparkles will be seen as the flour particles catch fire. The particles must be fine, with a large proportion of their surfaces exposed to oxygen of the air, to produce this effect.

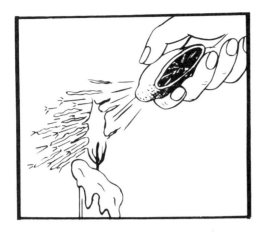

Fireworks from a Lemon

FLIP THE PENNY

NEEDED: A one-cent piece.

EXPERIMENT: Place the coin as shown in the upper photo, on the knuckle of the little finger. Flip the end of the finger with the thumb as shown. After a little practice, the coin can be made to turn over.

REASON: The upward movement of the finger simply gives the coin a twirling toss into the air. The little finger is hinged just beyond the coin and the sudden rotary motion of the finger gives rotary motion to the coin. It is necessary to flip the finger downward with the thumb. The little finger must be on the thumbnail. If the positions of the thumb and finger are reversed, the trick will not work.

The scientific principles involved here include muscle elasticity, inertia, momentum, and gravity.

THE MAGIC PENCIL

EXPERIMENT: Move your finger toward a pencil and make the pencil roll away mysteriously without touching it.

REASON: This is a trick that employs a scientific principle or two. As the finger approaches the pencil, blow on it, and the breeze will roll the pencil away.

The force exerted by the air as it moves to the pencil makes the pencil move, while a psychological factor magicians call "misdirection" comes into play, making the onlooker try to connect the finger movement with the pencil motion, when it actually has nothing to do with it.

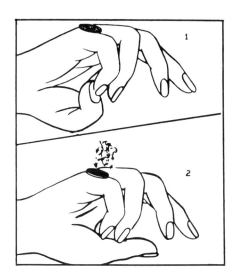

Flip the Penny

FAST MONEY

NEEDED: A dollar bill, preferably a crisp one.

EXPERIMENT: Hold the dollar bill as in drawing at upper left. Allow someone to hold his hand ready to catch it. As it is dropped it will move between the thumb and fingers before the lower hand can grasp it.

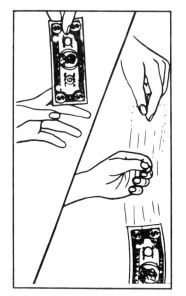

Fast Money

REASON: It takes time for the eye to see that the bill has been released, time for the brain to tell the hand to grasp, and still more time for the muscles in the hand to obey.

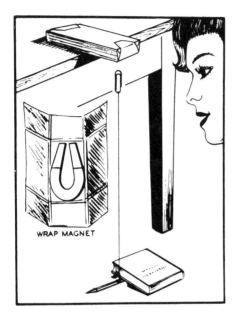

A Magnetic Mystery

WRAP MAGNET

A MAGNETIC MYSTERY

NEEDED: A magnet, paper, an iron paper clip, string, book, pencil.

EXPERIMENT: Wrap the magnet so that it is hidden; arrange the string and clip so that the book and pencil hold them down, and you have a mystery trick.

OBSERVATION: The magnet pulls against the iron in the paper clip, holding it up. The clip will stay suspended until it is moved far enough away from the magnet for gravity to exert a stronger pull than the magnet.

A BALLOON TRICK

NEEDED: A small rubber balloon, a drinking glass, some water.

EXPERIMENT: Hold the balloon in the glass as you blow it up. It expands against the glass and holds so tightly that the glass of water can be lifted by it.

REASON: The friction of the rubber against the glass may be sufficient to prevent its pulling out easily. But if the friction is not great enough, the balloon still will not pull out easily because air cannot get past it into the bottom of the glass. As it is pulled, the air pressure in the glass is reduced so that the greater pressure of the atmosphere on the bottom of the glass, exerted upward, can lift the weight of the glass of water.

THE SALT SHAKER MYSTERY

NEEDED: A salt shaker with salt and a steady tabletop.

EXPERIMENT: Place the shaker on a small pile of salt and balance it. Blow the salt away, and the shaker can remain balanced.

REASON: You do not blow *all* the salt away. The few grains needed to balance the shaker are held by it so that they do not blow away, although they may not be seen without very close examination.

This will require a little practice. If you blow too hard, the balance may be upset.

A Leaf Print

A LEAF PRINT

NEEDED: A warm crayon, a leaf, paper, a hot iron.

EXPERIMENT: Rub the back side of the leaf with the wax crayon until it is well covered with the wax. Place it between two sheets of paper, and press with a hot iron. The wax melts off of the leaf, making a pattern on the paper.

No explanation is necessary in this experiment, except that it is not necessary to place the hot iron on the leaf, since thin paper conducts the heat rapidly to the leaf, and the paper cannot smear the iron with any undesirable substance from the leaf.

The Ouija Board

THE OUIJA BOARD

NEEDED: A Ouija board and a partner.

EXPERIMENT: Place fingers lightly on the movable planchette. Do not rest arms or elbows on the table. Ask the Ouija a question and the planchette may move to spell out the answer.

REASON: The brain and the subconscious mind control the movements of the arm and finger muscles involuntarily. The answers are usually those the players want them to be.

Many people take the Ouija seriously, believing it as a link to the spirit world. But remember, if the spirits were moving the planchette, it would move without the fingers on it. Play Ouija as a game only.

An Optical Illusion

AN OPTICAL ILLUSION

NEEDED: Cardboard, scissors, protractor, pencil.

EXPERIMENT: Mark the card with the protractor, and cut it as shown. Write X on one piece, and O on the other. Place them so the O is above the X, and the X seems larger. Reverse the positions, and the O seems larger.

REASON: This is just another of the many optical illusions that may be made with simple materials. The brain takes into consideration not only what the eye sees at the moment, but the memories of what the eye saw before, and the meanings of those things the eye has seen before.

This is good; it enables us to observe more accurately, but sometimes it creates illusions, too.

Do You Dare?

44

DO YOU DARE?

NEEDED: A two-pound coffee can of sand, suspended overhead on a string, a "victim."

EXPERIMENT: Have the "victim" stand against the wall, pull the heavy can back until it touches his nose, then let it go (don't push.) See if he can stand still as the can swings back toward his nose.

REASON: According to the law of conservation of energy the can will never come back far enough to hit the "victim" on the nose. A little energy is always lost as the can swings. Of course, if the can is pushed instead of merely being released, it can come back far enough to hit the nose.

The author weighted the can with a brick instead of sand.

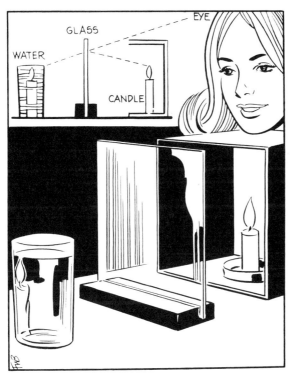

A Candle Burning in Water

A CANDLE BURNING IN WATER

NEEDED: Window glass, a glass of water, a burning candle in a can or box.

EXPERIMENT: Mount the glass in a slot in a board so it stands upright. Place the candle in the can in front of the glass as shown, so its reflection is seen in the glass. Stand the glass of water behind the window glass, place it correctly, and the candle will seem to be burning under the water.

REASON: The window glass reflects some light and lets some light pass through it. Light from the glass of water passes through to the eye, and at the same time some of the light from the candle is reflected from the window glass to the eye, creating the illusion. The drawing shows the paths of the light rays.

Chapter 5

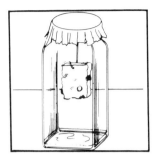

Biology & Psychology

CAN YOU HOLD A PENCIL?

NEEDED: A pencil.

EXPERIMENT: Hold the pencil (or other very light object even a feather will do) at arm's length for 20 minutes. You'll likely find it impossible to hold it even five minutes.

REASON: While the muscles of the arm are strong enough to lift and hold a much heavier object, they have not been trained and strengthened to support the weight of the hand and arm for a period of several minutes.

This, like many other physical feats, may be performed after considerable training and practice. A few *very strong* people may do this 20 minutes or more the first time they try it.

GHOST LIGHT

NEEDED: A bare electric bulb in a dark room.

EXPERIMENT: Turn on the light, gaze at it for ten seconds, then switch it off. You will continue to "see" light.

REASON: While the workings of the eye have not been satisfactorily explained, we know that the eye continues to see an object for a short period after the object is actually no longer visible.

This is called "persistence of vision." It is what allows us to see constant moving pictures at motion picture theaters or on our television screens, although these pictures are constantly flickering off and on.

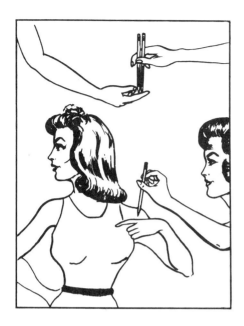

Sense of Touch

SENSE OF TOUCH

NEEDED: Two pencils and a partner.

EXPERIMENT: Touch the two pencils to the finger as shown, and the partner can feel both points easily. Try touching the partner on other parts of the body, and it will be difficult to tell whether one or two pencils are used.

REASON: Nerve endings that give us our sense of touch are close together in the fingertips but farther apart in most other parts of the body. Try the cheek, the tongue, the back, the arm, the leg, and the back of the hand.

It will be interesting to touch the pencil points close together, then repeat at the same place on the body with the points farther apart. See how far they must be in order that the partner can feel two separate pencils.

TWO LEAVES ALIKE

NEEDED: Leaves and marbles.

EXPERIMENT: Try to find two leaves exactly alike. It is impossible.

REASON: Nothing visible in living nature is exactly like anything else. In leaves, even if the tops seem alike, differences in the vein patterns may be seen. Crystal patterns of the same substances may be identical as well as molecules and smaller particles which we consider here as invisible.

To illustrate how impossible it is that collections of molecules or cells in living organisms could have the same configuration, place marbles together on the living room rug, and push another marble into the collection. Many of them will move; and it would be impossible to roll a marble against one group a second time and make the members move in the same way and form the same pattern.

WATER FROM LEAVES

NEEDED: A live plant, a sheet of cellophane.

EXPERIMENT: Wrap the plant with cellophane, tying it tightly around the stem. Droplets of moisture will form inside the cellophane.

REASON: A plant takes in water through its roots and gives off water through the leaves, flowers, and stems but mostly through the leaves. Not all the water taken in is given off; some of it is used by the plant in the manufacture of its food.

A potted plant or one growing in the ground is better, but the experiment may be performed with a stem cut and placed in water.

MUSCLE HABIT

EXPERIMENT: Try patting your head with both hands at the same time, and it is easy. Try rubbing your stomach with one hand and patting your head with the other, and it is quite difficult.

REASON: Automatic back-and-forth motions or circular motions of the hands and arms are performed easily, because the nervous system has been adapted to them through much repetition. Any change from the pattern formed is difficult and must be controlled by active thought in the brain. If your power of concentration is sufficiently strong, it can overcome the habit pattern and these motions can be performed.

WHY GET DIZZY?

NEEDED: Only yourself.

EXPERIMENT: Turn around rapidly, on foot or on a revolving stool, and soon you are dizzy.

REASON: Why? The cause is not completely understood. It is thought, however, that fluid in the semicircular canals of the ear begins to move around as we turn, tiny calcium compound particles in it brushing against tiny projections and registering the turning

Why Get Dizzy?

motion in the brain. When we stop turning, the fluid continues to turn, and the brain interprets this as if we are still in motion ourselves. The motion of the fluid soon stops after we stop, and we are normal again.

HOT OR COLD?

NEEDED: A friend, a match, an ice cube.

Hot or Cold?

EXPERIMENT 1: Have the friend cross the fingers of one hand, behind the back. Touch the crossed fingers with one finger, and the friend will believe two fingers were used.

EXPERIMENT 2: Tell the friend "I will blow out the match flame, and will touch you with the match. See if it still feels hot to you." Instead, touch the friend with the ice cube. The friend is likely to believe heat, not cold, is felt.

REASON: Our sense of touch operates very much by habit. Our fingers are not accustomed to the crossed position, and the touch of one finger registers as two.

Our nerves that sense heat and cold can be confused, and, if we are expecting heat, the cold ice cube may register in the brain as heat.

The Rising Arms

THE RISING ARMS

NEEDED: A doorway.

EXPERIMENT: Stand in the doorway, press outward with the hands and arms as shown in the drawing at right, and count slowly to 25. Step away from the door, and the arms begin to rise myteriously.

REASON: This is an interesting example of the workings of mind and muscle. The 25-count effort is sufficient to produce a persistent attempt to raise the arms. The doorcase prevents this, but as soon as you step out of the doorway, the persistent effort to raise the arms becomes a possible reality.

A TRICK OF THE MIND

NEEDED: Two opaque jars or cans and enough heavy metal to fill them.

EXPERIMENT 1: Fill one of the jars with something heavy, and screw on the lid so that the contents cannot be seen. Both jars should look alike. Have someone approach the table quickly and pick up the jars, one in either hand. The empty one will go up; the heavy one will be hardly raised from the table. (The person performing the experiment should not know that one is heavy.)

EXPERIMENT 2: If there is access to a chemical lab, find a container of mercury and ask someone to hand it to you. There may be great surprise when the friend finds he does not lift the container on the first try. The mind simply is not prepared to find it so heavy in proportion to its size.

REASON: Our minds direct the actions of our muscles, usually without our thinking about it. The mind's eye sees the two similar jars, and directs the muscle power it should take to raise them up, but, since they look alike, the muscles in both arms are commanded to lift equal weights. It takes an interval of time for the command to the muscles to be made when the mind learns that the jars are not equally heavy.

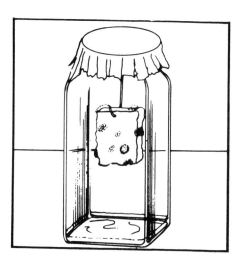

Bread Mold

BREAD MOLD

NEEDED: Glass jar or dish, foil, wire, water, a piece of bread.

EXPERIMENT 1: Hang the bread on the wire inside the jar, as shown, or simply place it on a dish. Put water in the jar or dish to keep the bread moist. The jar may be covered with foil to keep the moisture in. Mold will eventually appear on the bread.

EXPERIMENT 2: Examine some of the growths under a microscope or magnifying glass. Note their interesting shapes.

REASON: Fungus spores, including those that produce bread mold, are always in the air. They fall on the bread, and when conditions are right, they grow into mold. It may take several days for a good covering of mold to appear.

The mold will be in several colors. A blue-green kind which is likely to appear is penicillium, a member of the genus from which penicillin is extracted.

A MOLD-CULTURE MEDIUM

NEEDED: Potatoes, gelatin, a cooking pan, a large dish.

EXPERIMENT 1: Boil the potatoes until they disintegrate. Add one-fourth as much gelatin by volume, and pour into the dish to gell.

EXPERIMENT 2: Set a dish of the culture in a vessel of water and boil the water five minutes. Cover the culture with plastic and set it aside. Heat will probably have killed the tiny plants, and none will grow, at least for some time.

OBSERVATION: If the dish is allowed to set overnight at room temperature, mold will begin to grow. There will be spots in several colors, and the whole may become very attractive as the various kinds of molds spread to cover the dish. Leave the dish uncovered for this experiment.

REASON: The air always contains mold spores, as well as other microorganisms. The spores settle on the culture medium and begin to grow. Examine their structure under a magnifying glass. They are plants, but do not manufacture their food. They must get their food from the culture, or from living or decaying animal or vegetable matter.

GROW A FUNGUS GARDEN

NEEDED: A dish of culture medium, a piece of molded bread, a wire, a source of heat.

Grow a Fungus Garden

EXPERIMENT: Heat the wire to kill spores that may be clinging to it. Let the wire cool, and use it to transfer various colors of molds in any pattern desired. Heat and cool the wire each time a different kind of mold is touched with it. Cover the medium to keep out unwanted spores.

Try transplanting molds from blue cheese and other substances. Be sure all pieces of glassware are sterile before use. A handle on the wire will keep it from burning the fingers.

OBSERVATION: The culture medium will be covered in a few days with a beautiful "garden" of vari-colored molds. Molds are forms of fungi, plants that manufacture no food for themselves, but live as parasites.

(Experiment suggested by Al DeAnda, of the Mycology Lab, White Sands Missile Range.)

TREE FORCE

NEEDED: Observation.

The roots of a growing tree can exert unbelievable force. They can destroy concrete foundations and create massive upheavals of earth and rock. They can split a hard rock if growing inside it.

In the drawing a tree growing in a front yard is seen destroying concrete steps, the walk, and the street curb.

Tree Force

The large force exhibited by the roots is the combined forces of millions of tiny cells with fragile walls.

But there is very little energy involved. Energy or work equals force times distance moved. Power equals energy over time. Distance moved is small, energy small, time long (large), power small.

An Illusion

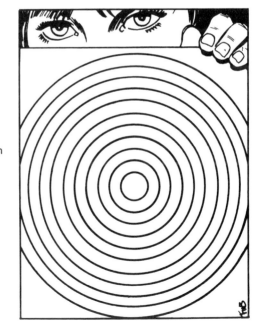

AN ILLUSION

NEEDED: This drawing.

EXPERIMENT: Make short, swift, motions with the drawing of circles, with the center of the circles moving in a small circle. The drawing will seem to turn like a wheel.

REASON: The apparent rotation of the circles is due to persistence of vision. This means that the eye sees the image after the visual stimulus ends, that is, after the circles have moved to another point.

This persistence of vision is useful to us when watching a movie or television. The pictures there do not remain constant, they go off and on. Yet our eyes do not turn them off when they actually go off, and so we get the impression of a constant but moving picture.

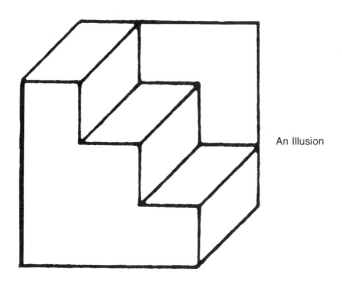

An Illusion

AN ILLUSION

NEEDED: The drawing shown here or a larger one like it.

EXPERIMENT: Look at the drawing of the staircase. Is it right side up or upside down? Look again and it likely will be reversed.

COMMENT: This is an old illusion known as the Schroder Staircase. It is called an equivocal figure and illustrates fluctuations of the vision process.

An Indoor Greenhouse

AN INDOOR GREENHOUSE

NEEDED: A pan or shallow box, coat hanger wire, plastic bag.

EXPERIMENT 1: Bend the wire so it makes a loop over the box as shown. Pull a plastic bag over it and tie the ends. Pull the excess under and fasten it with tape. This makes a greenhouse that can be placed in a window for starting seeds or cuttings.

The author made a box and drilled holes in the ends to hold the wire.

EXPERIMENT 2: Since the year 1909, when Dr. Robert Wood published an article in the *Philosophical Magazine*, many scientists have declared the "greenhouse effect" is false. They claim the heat build-up comes from the confining of the air, not on trapped heat vibrations. See if an experiment can be devised to prove whether this is correct.

A CRICKET THERMOMETER

NEEDED: A chirping cricket.

EXPERIMENT 1: Count the number of times your cricket

A Cricket Clock
Thermometer

chirps in 15 seconds, add 40, and you have the approximate temperature in degrees Fahrenheit.

REASON: Activities in many animals vary as temperatures vary. Snakes are a good example. In cold weather they are sluggish; in warm weather they slither along at high speed.

The cricket's voice is pleasant to most people, and the little fellow usually does no harm. If he finds a place of some seclusion near a fireplace, he may live there all winter.

EXPERIMENT 2: Crickets may be kept as pets. Put grass sod in a box, cover the box with mosquito netting, and insert the crickets. Keep the grass moist, and feed the crickets bread scraps and fruit rinds.

HOT GRASS

NEEDED: A bushel or more of fresh green grass clippings, crumpled newspapers, two thermometers.

EXPERIMENT 1: Make a pile or basket of crumpled paper the same size as the grass pile. Insert thermometers into each and keep track of the temperatures. At first they are likely to be the same, but by the next day the temperature of the grass may be 20 to 25 degrees higher, day and night, than that of the paper.

EXPERIMENT 2: If there is a cow or horse barn nearby, look at a pile of manure on a cold day. Vapor may be seen rising from it.

Hot Grass

The bacteria in it produce the heat necessary to drive water out in the form of vapor.

REASON: Bacteria begins to work at once in the clipped grass. They are alive, and produce heat as they eat, grow, and multiply. Sometimes a compost pile will get hot to the hand. It can catch fire from the heat it produces.

Fresh damp hay stored in a barn can catch fire and burn the hay and barn.

THE WORLD'S LARGEST MANUFACTURER

NEEDED: Houseplants.

EXPERIMENT: Water and give them plant food. Watch them grow!

COMMENT: The world's largest manufacturing process takes place in the cells of green plants. The process is called photosynthesis, and is necessary for the continuance of life on the earth.

With light as the energy source and catalyst, the plant takes up water and minerals from the soil, takes carbon dioxide from the air, and makes the food for the world. It is said that leaves use up 100 billion tons of carbon per year, transforming it into life-supporting forms.

World's Largest Manufacturer

According to the *New York Times* (11-13-66) all the blast furnaces in the world make only a half-billion tons of steel in the same period of time.

The first step in food synthesis may be summarized as: carbon dioxide plus water, with the aid of chlorophyll, yields a form of sugar, $C_6H_{12}O_6$ plus oxygen. Minerals are not involved in this step.

FREEZING OF TISSUE

NEEDED: Vegetables such as lettuce, cabbage, celery, carrots.

EXPERIMENT 1: Place the vegetables in a freezer and look at them every few minutes. Some will freeze more quickly than others.

EXPERIMENT 2: To show this with water, place a little water in two drinking glasses. Put a little salt in one, then set both in the freezer. The water without the salt will freeze first.

Freezing of Tissue

REASON: Lettuce froze quickest for the author, perhaps because the water in it contained a smaller concentration of chemicals and because it offered more surface cells from which to lose heat. Other vegetables froze at different times; the less pure the water the lower their freezing points.

A PINCH FOR PAIN

NEEDED: Someone with leg cramps or other leg pain.

EXPERIMENT 1: Have the person pinch the upper lip as shown in the drawing, and see if the pain vanishes.

EXPERIMENT 2: To suppress a sneeze, press the finger firmly along the upper lip. The sneeze coming on probably will be stopped before it develops. Try it!

REASON: Milton F. Allen, businessman of P.O. Box 789, Decatur, Georgia suffered severe leg cramps, and discovered that such a pinch of the face above the lip relieved the pain. He calls it "acupinch."

Mr. Allen's discovery has been publicized, and he has received hundreds of letters from people who found acupinch a successful treatment of their cases.

Mr. Allen, a religious man rather than a scientist, credits his discovery to prayer he offered when the pain was almost unbearable. He offers no explanation as to why a pinch of the face cures

A Pinch for Pain

Minerals from the Earth

62

pain in a leg. He says it must be akin to acupuncture. This may not be profound science, but is harmless and interesting to try.

MINERALS FROM THE EARTH

NEEDED: Flower pot, small stones, soil, water, a vessel to catch the water.

EXPERIMENT 1: Place some stones in the bottom of the pot, and soil above them. Place the pot into or above the container, and pour water into the pot. As it dribbles through the soil it takes up many nutrients that will be helpful in making plants grow.

EXPERIMENT 2: Try using three plants. Put one in garden soil, one in sand, and another in garden soil. Water the first with your enriched soil water. Water the others with distilled water. Compare their growth.

REASON: Plant nutrients must be water soluble, that is, capable of being dissolved in water. Many of them dissolve in the water as it seeps through the soil in the pot.

Try watering house plants with this "enriched" water. They should grow better than plants watered with regular tap water.

New Plants from Leaves

NEW PLANTS FROM LEAVES

NEEDED: African violet plant, potting soil, Rootone, water, a clay pot.

EXPERIMENT: Break a mature leaf from the plant, dip the broken end in Rootone, and place the stem in water or in soil. In two or four weeks a new plant will be growing from the broken stem. Cut or break off the new plant—and the same leaf and stem may be used again to start a new plant.

COMMENT: If the leaf is placed in soil put a plastic bag over the pot to hold moisture. Keep away from direct sunlight. If the city water contains chlorine (and it probably does) it is well to let water set overnight in a plastic or enamel container, then heat it slightly and let it cool before using it to water a plant. This will take most of the chlorine out.

See the Pulse Beat

SEE THE PULSE BEAT

NEEDED: A kitchen match, a dime, a candle, a table.

EXPERIMENT 1: Drop wax from the candle on the dime and stand the match in it so the match stands upright. Place the dime on the wrist, and hold the arm and hand still on the table. If the dime is on the right place the match will be seen to move back and forth slightly as the pulse beats.

EXPERIMENT 2: Look for parts of the body where the motion of blood vessels indicate heartbeats.

EXPERIMENT 3: Have a friend exercise, then notice how much stronger the beats show.

REASON: The heart pumps blood through the body in an uneven manner. This can be observed better in some parts of the body than in others. The wrist is a good place to notice the heartthrobs because the blood vessels are near the surface of the skin.

All Natural Dyes

ALL NATURAL DYES

NEEDED: A beet, a carrot, a squash, beans, cabbage.

EXPERIMENT 1: Boil the vegetables separately. The water in which the beets are boiled will be red; water in which the other vegetables are boiled will remain almost clear.

EXPERIMENT 2: Natural dyes were once used for dyeing all fabrics for our ancestors' clothing. Walnut hulls would give brown dye; brown onion skin, light brown; hickory bark or chips, yellow;

green grass, green; red clay, various shades of red. Boil these substances in water to obtain dyes from them.

REASON: Natural dyes give vegetables their color, and these dyes are different in several ways. The natural dye in beets, for example, is readily soluble in water while the dyes in the other vegetables are not readily soluble in hot water.

The Absorbing Sand

THE ABSORBING SAND

NEEDED: Three cans, one filled with red clay, another with garden topsoil, another filled with sand, water.

EXPERIMENT: Pack the sand and soil down and pour an equal amount of water on top of each. Watch what happens.

OBSERVATION: The water poured into the sand will pass through it rapidly. Water poured into the clay will stand on top of it a long time. Water poured into the topsoil will seep into it slowly and be held there.

REASON: When the earth is stripped down to the clay, the rain that falls on it will run off very fast, and possibly produce flooding. Sandy soil will let the water pass down to harder and more compact layers, possibly too deep to support plant life well. Topsoil holds rain, making it available to plants for good growth.

Once good topsoil is destroyed it cannot be rebuilt easily and quickly. This is good for ecology-minded people to remember.

VALVES IN THE ARM

NEEDED: Someone with prominent veins in the arm.

EXPERIMENT: Rub a vein backward, from the heart, and watch little knots form at various places on the vein. These are

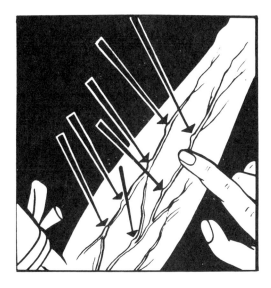

Valves in the Arm

valves which let the blood flow only in the right direction, through the veins to the heart. Sometimes a tourniquet will make the valves show up. Squeezing the fist helps, too.

COMMENT: Back in 1628 William Harvey used these little demonstrations in proving that the blood circulates. He published

Tasteless Coffee

his findings in the book "On the Motion of the Heart and Blood in Animals" which may be seen in a library. He upset several of the superstitions about the body, superstitions that were accepted as fact at the time.

TASTELESS COFFEE

NEEDED: Some dry coffee.

EXPERIMENT: Hold your nose and chew some dry coffee—or let some instant coffee dissolve on the tongue. The good coffee taste is not there. Take a breath, and the taste is there.

REASON: The sense of taste is limited to a few basic tastes, and unless a substance affects these there is no taste apparent. But the sense of smell is one of our wonderfully all-embracing senses. It can detect hundreds of different odors.

So closely related are these senses that it is difficult sometimes to determine which we are experiencing. The "taste" of many foods is not taste at all, but odor. The coffee is an example.

This explains why foods do not taste right when we have a cold. The odors cannot get through clogged sinuses to the sensors of smell.

Chapter 6

Water & Surface Tension

HOLES IN THE WATER

NEEDED: A pan of water, four identical blocks of wood, wax paper.

EXPERIMENT: Wrap two of the blocks with wax paper

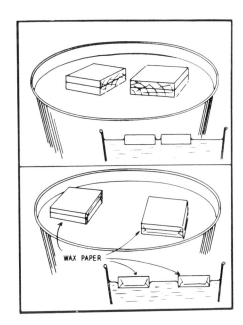

Holes in the Water

(thumbtacks may be used to hold the paper around the blocks). Place the two unwrapped blocks close together on the water, and they will come together. Place the wrapped blocks on the water. They do not come together; they may push farther apart.

REASON: Notice that the water wets and clings to the unwrapped blocks as in the diagram, actually pulling itself up along the sides of the blocks. The surface tension of the water between the blocks, acting like a stretched rubber sheet, pulls the blocks together.

The wax paper is not "wetted" by the water; it seems to cause the weight of the blocks to actually make holes in the water. The lower diagram shows how this looks on close observation.

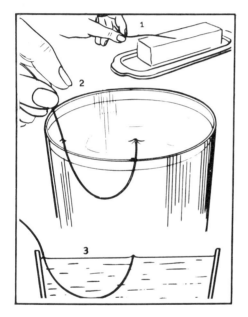

Stretch a Water Surface

STRETCH A WATER SURFACE

NEEDED: A glass of water and a small stiff wire.

EXPERIMENT: Place the wire under the water surface and bring it up slowly as in the large drawing. If care is taken, and the wire moved very slowly, the surface of the water may be seen to stretch upward before the wire finally breaks through it.

REASON: The surface tension of water is much like a stretched rubber sheet. It actually takes a little force to break through it, just as it would take a greater force to pierce a rubber sheet.

If the experiment does not work, try greasing the end of the wire with butter or oil.

SOAPY SMEAR

NEEDED: Pieces of flat glass, soapy water, plain water, two atomizers.

EXPERIMENT: Spray the soapy water on one piece of glass, the plain water on the other. The plain water will form tiny drops; the soapy water will run down more readily and drip off the glass.

REASON: Surface tension tends to make a liquid assume the shape which has the smallest surface area, which is the sphere. This is seen when the plain water is sprayed onto the glass.

Soap or detergent lowers the surface tension, and gravity pulls the liquid downward as a thin film.

Wetter Water

WETTER WATER

NEEDED: Water, talcum powder, liquid detergent, a needle.

EXPERIMENT 1: Lower the needle gently to the surface of the water. It can be made to float there. Pour a little detergent on the water, and as it reaches the needle, the needle sinks.

EXPERIMENT 2: Clean the container well, put water into it again, and sprinkle talcum powder on the surface. Pour on some detergent, and it will make a path through the powder as shown.

EXPERIMENT 3: Clean the container again. Put powder and the needle both on the water surface. Watch them sink as detergent reaches them together.

REASON: The surface tension, a surface film on the water, is strong enough to support the needle. The detergent breaks the film.

More Experiments. Use pepper instead of talcum powder. Use soap instead of regular detergent. (Soap is a detergent, too, but is weaker than regular detergents from the store.)

OILY SMEAR

NEEDED: Window glass, oil, water, a spatula.

EXPERIMENT: Place only a touch of an oil drop on the glass, also several drops of water. Smear with the flat edge of the spatula, and the fogginess of the windshield wiper in a light rain can be duplicated.

REASON: On the windshield, the small amount of oil usually comes from drops of water tossed up from the highway by tires of other vehicles—rarely from the wiping cloths at filling stations.

Water on an oily surface forms into tiny round balls if there is a very small amount of it. This is because surface tension tends to make the water droplets take the shape of spheres. When the rubber spatula or blade is rubbed over the water it simply divides the drops into many smaller droplets.

The fogginess clears as the blade moves along, because the amount of water involved is so small that it evaporates quickly.

THE SWEATING JARS

NEEDED: Two jars, ice water, hot water.

EXPERIMENT 1: Put hot water into a jar, filling it half full. Note that droplets of water appear on the inside of the glass above the water.

EXPERIMENT 2: Put ice water into a jar, filling it half full. Note that droplets of water form on the lower half of the jar, where the ice water cools it below room temperature.

REASON: Vapor from the hot water rises in the jar, and as it comes into contact with the cooler parts of the glass some of it condenses out and sticks to the glass in droplets.

Air can hold more moisture when warm than when it is cold. As warm air from the room touches the cold glass and becomes cooler, it cannot hold as much moisture and must give up some. Part of what it gives up forms droplets on the glass.

The Mysterious Needle

The same principle applies to both of these experiments.

THE MYSTERIOUS NEEDLE

NEEDED: Wire, pliers, some strong suds.

EXPERIMENT: Make a wire loop with a handle, and cut another piece of wire to rest across it. The piece should be very straight, and all the wires should be clean and smooth.

Place the short piece on the loop, dip both into the suds, and lift them out. There should be a water film covering the entire loop. Break the film on one side of the wire with the finger, and the remaining film will draw the wire toward the edge of the loop.

REASON: The surface of ordinary or soapy water acts as a stretched rubber sheet. This is surface tension, and it is strong enough to pull the small wire or needle along.

EROSION OF SOIL

NEEDED: Soil on a board, an equal area of grass sod on a board, a water hose.

EXPERIMENT 1: Spray water on the soil, and it washes into gullies and may all wash away. Spray the sod the same way, and it resists the washing action of the water.

EXPERIMENT 2: Notice that farmers make use of "contour

Erosion of Soil

plowing" to prevent washing of the soil. This means that plowing follows the contours of the land, never allowing rain to wash directly downhill in a plowed furrow.

An experiment can be made to show this if a large square of sod is available. Arrange it so that low places follow a contour, while in another section allow the low "furrows" to run up and down the hill. Water will be seen to flow down much more slowly on the contoured sod.

REASON: The grass softens the blows hit by the drops of water, and the tangle of roots softens the flow of water through the soil underneath. This shows why soil stripped of its grasses or trees soon washes away.

WACKY WICK

NEEDED: Two jars, one filled with water, a cloth, a wire.

EXPERIMENT 1: Roll the cloth around the wire. The wire is to give it stiffness. Bend it. The water will move through the cloth from the upper to the lower jar.

EXPERIMENT 2: This siphon can be used as a filter. Put slightly muddy water in the upper jar, and as it siphons over into the lower jar most of the mud is left behind in the cloth. This is because the particles of mud do not flow through the cloth as does the clear water.

While this is a filter of sorts, it does not purify the water enough for drinking. Look, but don't drink.

REASON: Capillary action, in which the water molecules cling to the tiny fibers of the cloth, causes the water to rise into the cloth against the force of gravity. When the water gets over the edge of the glass, however, both gravity and capillarity combine to pull it downward into the second glass.

"Capillary" comes from a Latin word which means hair, and refers to the tiny bores in "capillary tubes." These are tubes in which water will rise against the force of gravity, the height of the rise depending on the smallness of the bore.

MELTING UNDER PRESSURE

NEEDED: Two ice cubes

EXPERIMENT: Squeeze the ice cubes together. Some of the ice will melt under the pressure, then should freeze again when the pressure is removed, sticking the cubes together.

REASON: Notice that *dry* ice and snow are not too slippery. Actually, the skater skates on water, because the pressure of the steel blade on the ice melts a little of the ice, to provide the slippery water layer. Of course, the water ordinarily freezes again when the skater has gone and the pressure is removed, if air temperature is below the freezing point of water.

MARRIAGE OF THE WATER DROPS

NEEDED: A piece of clean glass and some water drops.

EXPERIMENT: Place drops close together on the glass. They will move closer until they join together.

REASON: Molecules of water attract one another. Gravity flattens the drops on the glass until they touch each other; surface tension then pulls them to a common center of mass. The surface tension, which acts something like a stretched rubber sheet, combines the separate drops so that they have a shape with the least possible surface.

The surface tension is greater than the force of cohesion between the water and the glass.

Dirty glass would reduce the surface tension of the water.

AN OIL-DROP ENGINE

NEEDED: A pan of still water, a piece of cardboard cut as shown, a drop of oil.

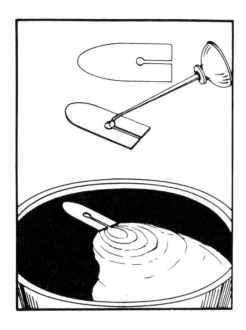

An Oil-Drop Engine

EXPERIMENT: Place the boat on the water. Place a drop of oil carefully in the hole. The boat will move forward.

REASON: The oil, which is lighter than the water, floats on the surface. As it runs out the rear opening, the surface tension is reduced and the surface tension at the front is not reduced. The boat moves in the direction of the greater pull.

Could this be an example of "action-reaction" also? If the oil flows out the back, reaction would tend to propel the boat forward. The rule is that for every action there is an equal and opposite reaction.

WHY THE ROUND DROPS?

NEEDED: A glass of water, rubbing alcohol, castor oil, a medicine dropper.

EXPERIMENT 1: Mix water and alcohol until it is the same weight as the oil. A little experimentation will show when the right proportions of water and alcohol are reached. If the drops rise, add more alcohol. If they move downward, add more water. Drops of oil then, when placed under the surface, remain in their places and assume a round shape.

EXPERIMENT 2: Use a clean glass or jar. Place water in it, and let the alcohol down slowly over the water so they do not mix

Why the Round Drops?

very much. The oil may then be placed between water and alcohol, and will float there.

REASON: The surface tension of the oil pulls the drops into the shape which has the smallest surface area, which is the sphere.

MERGING STREAMS

NEEDED: A tin can, a nail, hammer, water.

EXPERIMENT: Make three holes in the can near the bottom. They should be about 3/16 of an inch apart. Fill the can with water, and three streams will come out. If the streams are pinched together they merge into one. Cover the middle hole momentarily with the finger, and three streams will form again.

REASON: The molecules of water cling together—cohesion. When the streams are united, surface tension also comes into the picture. This is like a film stretched over a water surface, and the streams do not break through it easily.

THE LIVELY SOAP

NEEDED: A dish of clean water and a tiny piece of soap.

EXPERIMENT: Use a piece of soap half the size of a pinhead. Drop it on the water, watch carefully, and the soap will move around quite fast.

REASON: The surface tension of the water acts like a stretched rubber sheet. The soap reduces the tension, but is not likely to do so evenly. The tension pulls the piece along on the side where it has acted least to affect the surface tension. The movement stops when there is an equal amount of soap over all the surface.

A Cloth and Sponge Mystery

A CLOTH AND SPONGE MYSTERY

NEEDED: A pan of water, a sponge, a strip of cotton cloth.

EXPERIMENT 1: Place the sponge in the water, with the cloth folded on top of it. The water will rise rather quickly to the top of the sponge, but much more slowly through the cloth.

EXPERIMENT 2: Try placing another sponge above the first in place of the cloth. Will the water rise into the second sponge more quickly than it did through the cloth?

EXPERIMENT 3: Try using dry sponges, placing one in the water and the other above it, but this time place a paperweight or other small weight on the second sponge. This should put the sponges into closer contact with each other, allowing the water to flow more easily into the top sponge.

REASON: In the sponge, the fibers are close together, so that the water can rise easily through the sponge by the process known as capillary action. There is no close network of fibers linking the folds of cloth, and so the water practically stops there. Capillary action allows oil to rise in the wick of a lamp, or water to move up through the ground.

WHAT HOLDS UP THE CLOUDS?

NEEDED: A milk bottle, some hot water, an ice cube.

EXPERIMENT: Put hot water into the bottle, place the ice cube over the mouth, and small clouds will be seen to form inside the bottle. What holds them up?

Actually, if left still and not disturbed by air currents, all clouds would eventually fall in air at a constant temperature. But the droplets of water in clouds are so small and light that they do not fall rapidly because of the air resistance. The slightest air current can blow them along. If they join together and become heavier they fall faster, usually as rain.

Rising currents of warm air give an upward push to the tiny droplets and the clouds move up or down according to the amount of the upward-moving air. Watch fog carefully, and you can often see it fall.

HOW PURE IS THE WATER?

NEEDED: A clean glass filled with water.

EXPERIMENT: Let the glass stand on a shelf until the water has evaporated. Notice that the glass will not be clean.

REASON: Ordinary water contains many substances in a dissolved or suspended state. Most of them do not evaporate, so, as the water evaporates or goes into the air as vapor, it leaves the other substances behind. They cling to the glass sides and bottom.

This is not an accurate check on impurities in the water, since some of the residue seen as the water leaves the glass consists of substances that have settled in the water from the air as the glass stood on the shelf.

Such "impurities" as we find by this method do not adversely affect the water as we drink it or otherwise use it. Many of them are good for health. Special tests are needed if water is thought to be unsafe and these tests are not performed at home.

A WATER AND WEIGHT MYSTERY

NEEDED: A container of water, a scale for weighing, a block of wood.

EXPERIMENT 1: Place the empty container on the scale, weigh it, then put the block of wood into it. The additional weight registered is that of the wood. Take the block out.

EXPERIMENT 2: Put water in the container, note the weight, and again add the wood. Again the added weight is that of the wood.

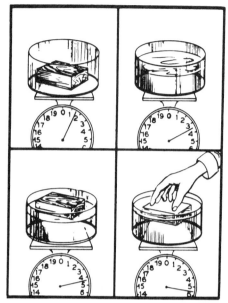

A Water and Weight Mystery

EXPERIMENT 3: Press the wood down with the fingertips until it is barely under the water. This time the added weight is more than that of the wood. Why?

REASON: Archimedes discovered more than 2,000 years ago that an object immersed in water but not touching bottom adds weight equal to that of the water displaced. When the wood is pushed down by force from the fingertips until it is under water more water is displaced.

SHIMMERING LIQUIDS

NEEDED: Water in a flat pan, some rubbing alcohol.

EXPERIMENT: Pour a little of the alcohol into the center of the pan of water. A beautiful shimmering effect is seen moving outward toward the edges of the pan. The shimmering continues until the liquids are mixed.

REASON: Each liquid when separate has its different surface tension. As the alcohol spreads outward, the surface tension of the alcohol-water surface becomes less than the surface tension of the water, so that there is an unbalanced pull. This causes the slight outward shimmering motion which continues until the surface tension pulls are equalized.

WATER ON THE WALL

EXPERIMENT: Notice how water drops form on the wall of the bathroom when you take a shower.

REASON: Warm air will hold more water than colder air. The warm water in the shower warms the air and at the same time offers an ideal condition for evaporation of water into the air.

When this same air, with its load of water, meets the cooler walls of the room, its temperature is lowered. Since it can then hold less water than when it was warm, some of its water is condensed against the wall.

EVAPORATION AND SWEATING

NEEDED: Observation.

EXPERIMENT: Open the bathroom door after taking a shower. The air is cold until the body is dry.

REASON: Water, to evaporate, takes energy. Evaporating from the body, water takes energy mainly in the form of body heat. The body actually becomes cooler during the evaporation process. This is one of the reasons we sweat on a hot day.

Water evaporates more quickly from a human body than it does from a table in a 70 degree room because the water on the body is nearer its boiling point.

Stubborn Paper

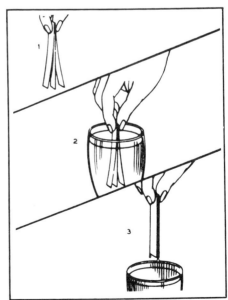

STUBBORN PAPER

NEEDED: Three strips of paper held together as shown in the drawing, a glass of water.

EXPERIMENT 1: Notice that while the papers are dry they stand apart as shown on the previous page. In the water they still stand apart, but when they are brought out of the water they stubbornly cling together.

EXPERIMENT 2: A piece of wet paper towel sticks to the smooth top of the kitchen table: adhesion.

EXPERIMENT 3: See if a piece of waxed paper will stick to the smooth tabletop as does the paper towel. Wet the paper first; it will be difficult to wet. It may be necessary to put water on the table then press the paper into the water to make it stick.

REASON: When the strips are surrounded by large volumes of air or water there is little or no effect on them due to surface tension. The tension is the same on all sides of the papers when they are immersed. But when they are lifted from the water the water films remaining on them are pulled by their surface tension into the least possible space, and the least space is when the strips are made to cling together.

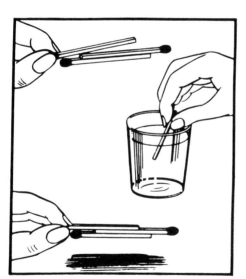

Stuck Sticks

STUCK STICKS

NEEDED: Three matches with square sticks, some water.
EXPERIMENT 1: Challenge a friend to lift two matches with

one. It is easy if the sticks are wet; the two matches may be lifted as shown in the drawing.

EXPERIMENT 2: Most grains of cereal will stick together when in milk. This is more adhesion.

REASON: Molecules of water cling together: this is called cohesion. They also cling to the match sticks: this is called adhesion. The adhesive force of the sticks and water is great enough to support the weight of the sticks.

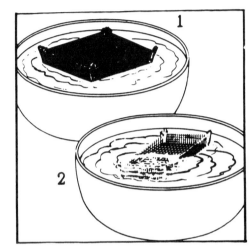

Boat of Holes

BOAT OF HOLES

NEEDED: A piece of new screen wire, a dish of water, some detergent.

EXPERIMENT 1: Bend the wire mesh into the form of a flat bottom boat. Place it carefully, flat, on the surface of water. It floats.

EXPERIMENT 2: Push a side or corner of the wire boat under, and the whole boat sinks rapidly. If part of the surface tension film is broken the weight of the boat causes it to break the films at other parts, and the boat sinks.

EXPERIMENT 3: Float the boat again, and pour a little detergent into the water. The detergent breaks the surface film and the boat sinks.

REASON: Surface tension on the water is like a stretched film, and is strong enough that the wires of the boat do not break through easily.

Water Stream from a Faucet

WATER STREAM FROM A FAUCET

NEEDED: Observation

EXPERIMENT: Turn on the water in the sink. If the flow is right the stream starts the size of the faucet opening, then grows smaller as it falls.

REASON: Surface tension tends to hold the water in a solid stream. But as it falls its speed increases because of the pull of gravity. It is in continuous flow, which means that the amount passing any point must be the same. If it is flowing faster the size must be reduced to meet this law of physics.

The amount of flow is the cross-sectional area of the stream times the speed of flow of the water.

CAPILLARITY

NEEDED: Strips of paper towel, salt water, soapy water, plain water, rubbing alcohol, oil.

EXPERIMENT: Dip the ends of the paper strips into the different liquids, and notice that the liquids do not move up into the paper, between the fibers, to the same extent.

REASON: "Sometimes the forces of cohesion and adhesion oppose each other; this brings about the effect known as capillarity." (The *Book of Popular Science* explanation).

Capillarity is a surface tension effect. The small openings in the paper towel are capillary openings. The molecules of the different materials and the dissolved materials in water have a varied effect because of variation in their surface tension.

THE FISHERMAN'S PUZZLE

NEEDED: A jar of water, a can, nails, a marking crayon.

Capillarity

EXPERIMENT: A fisherman in a boat on a very small lake throws some heavy iron out of the boat and into the water. Does the act raise or lower the level of the water in the lake?

The Fisherman's Puzzle

Let the jar represent the lake, the can the boat. Put nails in the can and float it on the water in the jar. Mark the water level. Now take the nails out of the can and drop them into the water. Replace the can so it floats again. Note that the water level in the jar is lower.

REASON: In the boat or the can the iron (nails usually are iron) displaces a weight of water equal to its weight because they are floating with the boat. In the water the iron displaces only an amount of water equal to its volume. Since the iron is much heavier than water its volume is much less than the volume of water equal to its weight.

Funnel Falling

FUNNEL FALLING

NEEDED: Dry sand, sugar, salt or other grains or crystals, funnel.

EXPERIMENT: Fill the funnel, then watch as the finger is removed and the grains "pour" out as if they were liquid.

COMMENT: A liquid can be described as a substance that, unlike a solid, flows readily, but unlike a gas, does not expand indefinitely. The grains or crystals fit this description.

The middle grains fall first, making a cone-shaped design like water. But, unlike water, they do not begin to swirl as they fall through the funnel.

Water Bullets

WATER BULLETS

NEEDED: A pan of water, a thermometer, a fan.

EXPERIMENT 1: Place water in the pan. We know it will evaporate if left alone, and will evaporate faster if air from the fan blows over it. Place the thermometer in the water, and it will be cooler when the fan blows air over it.

REASON: Molecules of the water are in constant motion, as are molecules in any fluid. Some of them gain a high enough speed to shoot like bullets out of the water into the air above. Many fall back into the water; some escape. More can escape if a breeze is blowing to blow them away.

EXPERIMENT 2: Why does hot water evaporate faster than cold water? The molecules move faster and more of them escape the solution.

Evaporation is an example of expansion, and expansion means cooling. In this case the water molecules escaping carry away some of the kinetic energy of the water, and a lowering of the kinetic energy means a lowering of the temperature.

AN EVAPORATION TRICK

NEEDED: Two similar jars, rubber from a large broken balloon, water, rubber bands.

EXPERIMENT: Fill one jar almost full of water. Set both jars

An Evaporation Trick

aside in a warm room for a few hours, until the temperature in each jar is the same. Cover both jars with stretched pieces of rubber held on tightly with rubber bands, sealing the jars.

The rubber over the jar of water will be seen to bulge slightly, showing that a little pressure has been generated in the small air space above the water.

REASON: Molecules of water, always moving, sometimes jump out of the water and into the air. This is evaporation, and can increase the pressure in the air enough to bulge the rubber. The jar of air has nothing to evaporate and so the pressure remains the same as long as the temperature in the jars remains the same.

The bulge is somewhat exaggerated in the drawing to make for clarity. The jars should remain several hours or overnight after being sealed with the rubber to show the bulge.

BLOWING SNOW

NEEDED: Observation of light snow in the wind.

EXPERIMENT 1: Observe the snow on a windy day. When the wind is strong the snow seems to blow away, while the temperature is below freezing and we know it cannot melt.

EXPERIMENT 2: Leave some mothballs out in the open. They will disappear or "sublime." This takes a little time.

REASON: Snow is made of ice crystals, and ice is one of the substances that can pass from the solid to the gaseous state without going through the liquid state. This is called *sublimation*.

Blowing Snow

So, we see that the snow sublimes as the wind blows it about—without melting. Large pieces of ice do not usually do this to a noticeable extent, however.

Chapter 7

Gravity & Centrifugal Force

A SAND AND WATER MYSTERY

NEEDED: Sand, water, a glass

EXPERIMENT: Stir sand and water in a glass, and the sand will be seen to form a pile in the center as shown.

REASON: Because of the friction between the glass and the

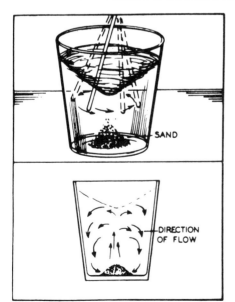

SAND

DIRECTION OF FLOW

A Sand and Water Mystery

water, the bottom section of the water moves more slowly; therefore less centrifugal force is exerted. This causes the water to flow downward along the sides of the glass, inward across the bottom.

The upward flow of the water is not sufficient to carry the sand upward along the complete water path, but just enough to pile it up on the bottom of the glass.

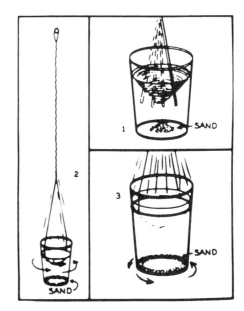

More Sand and Water Mystery

MORE SAND AND WATER MYSTERY

NEEDED: A glass of water, some sand, some string.

EXPERIMENT 1: Stir the water and sand together, and the sand will pile up in the middle of the glass as the water turns around.

EXPERIMENT 2: Suspend the glass by the string, twist the string and release it so that the glass with the water will whirl around. This time the sand will go outward along the bottom of the glass.

REASON: (1) Friction between water and glass at the bottom is greater than the friction higher up in the glass; therefore the water at the bottom turns more slowly. This causes downward currents along the sides of the glass, inward at the bottom, and upward in the center. The water motion carries the sand to the center.

(2) When the glass, too, turns with the water and sand, all turn at the same rate, so that there is no greater friction at the bottom

than higher up. Therefore, there is no flow of water as in (1). The sand can then follow its natural tendency to move off at a tangent to the circular motion, which means that it can move to the rim of the glass as shown.

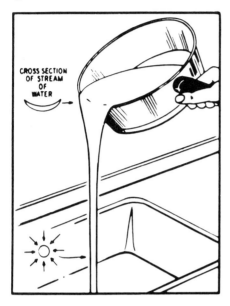

Shape of the Poured Water

SHAPE OF THE POURED WATER

NEEDED: A pan of water.

EXPERIMENT 1: Pour the water out of the pan, and notice that the stream becomes round as it flows down a few inches from the pan.

EXPERIMENT 2: Hold the pan higher, and the stream will tend to divide into drops.

REASON: Gravity gives the water in the pan a level surface, and as it pours out of the pan, it takes the form shown in the diagram at the top.

As it pours down, gravity and the shape of the pan influence it less and less, until surface tension pulls the stream into the shape shown in the lower cross-section drawing.

Surface tension is a force which tends to give water and other liquids the shape with the smallest surface area.

WHY THE VORTEX?

NEEDED: A round pan or jar, some water, something to stir with.

EXPERIMENT 1: Stir the water, and the faster it goes around the deeper will be the "hole" in the center. This hole with the rotating water around it is called a vortex.

EXPERIMENT 2: Use something larger to stir with, such as a spoon. The water may be made to reach the top of the container and flow over the rim.

REASON: Centrifugal force supplied by the stirrer tends to cause the water to fly away from the center. The wall of the vessel will not let the water fly off.

The pressure under the water increases with depth. At the bottom of the vortex the force due to pressure will not allow the water to move out near the side of the glass as it does near the top.

Chapter 8

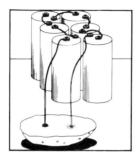

Electricity & Magnetism

A POTATO TEST

NEEDED: A battery, a potato, some wire.

EXPERIMENT 1: Connect wires to the battery. Cut the potato in half, and stick the ends of the wires into the potato about an inch apart. A green color will appear in the potato around the positive wire, and bubbles will come from the potato where the negative wire is inserted.

REASON: The green color is due to partial ionization of the copper as negative ions from the solution in the potato are neutralized at the anode pole and attack the copper to form an ionic copper salt. This action is very slight since the concentration of ions in the potato juice are very low. The bubbles of gas are due to liberation of tiny volumes of hydrogen gas at the negative pole.

This experiment has been suggested to determine which of the two wires is positive and which is negative.

Even if an inert wire, such as platinum, is used, the potato will still show color around the positive wire. The color will be a pale pink, or dirty pale pink. This is caused by oxidization of compounds in the potato by oxygen, or other oxidizing substances such as chlorine released by the electrochemical reaction.

It is the same reaction that causes a cut potato to brown if it is left in the air (apples and peaches will brown the same way if exposed to air or other oxidizing conditions after they are cut).

When preparing things that brown for deep freezing it is cus-

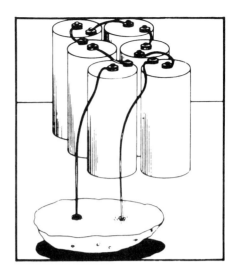

A Potato Test

tomary to sprinkle them with ascorbic acid, vitamin C, sold under a number of brand names, for the purpose of keeping them from turning brown. Vitamin C (harmless to the body) is a reducing agent; that is, it reacts with, and takes up, oxygen easier than the materials in the fruit which would turn brown on oxidation.

(Such oxidation destroys the vitamin properties of ascorbic acid, but it does protect the fruit from browning.) Lemon juice contains ascorbic acid and will protect a cut potato or fruit from browning in the air.

EXPERIMENT 2: Put a drop of lemon juice on the potato at the point where the positive wire enters and see if it will prevent the formation of a color at that point.

THE GOOFY PING-PONG BALL

NEEDED: A table tennis ball and a hard rubber comb.

EXPERIMENT: Rub the comb briskly on a woolen sleeve or cloth, move it around in circles quickly as shown, and the ball will follow it. It is not necessary to touch the ball with the comb.

REASON: Rubbing places a charge of static electricity on the comb. The uncharged ball is attracted by the charge on the comb in a very mysterious manner.

The rule is: like charges repel, unlike charges attract, and a charged object near an uncharged object induces an opposite charge on the near side of the formerly uncharged object. The induced charge is due to a shift of electrons on the surface of the formerly uncharged object.

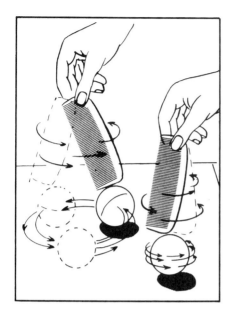

The Goofy Ping-Pong Ball

THE WAYWARD WATER

NEEDED: A hard rubber comb and a thin stream of water from a faucet.

EXPERIMENT: Rub the comb briskly on a woolen cloth or through the hair, then hold it near the stream of water. The water will change its course as it is attracted by the comb.

REASON: Rubbing the comb puts a charge of static electricity on it. The water is not charged until close proximity of the comb repels electrons to the opposite side of the water column. Then the charge is opposite, and the comb and water are attracted to each other. Note that the rule in electricity: "unlike charges attract one another" can be expanded to include attraction between a charged body and a previously uncharged body which has an opposite charge induced in it when close to a charged body. This experiment works only when the air is dry. It will not work in a room where the humidity is high. Much water vapor in air allows the excess electrons of the charged comb to escape from the comb to invisible water vapor aggregates in air.

MAGNETIC OR NOT?

NEEDED: A magnet, a tin can lid, various substances to be tested such as paper, cloth, screen wire, and aluminum.

EXPERIMENT: Lift the can lid with the magnet (it is not

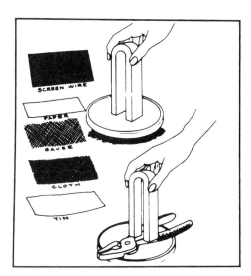

Magnetic or Not?

tin—it is iron), as shown in the upper drawing. Then try lifting it with various substances or objects placed between the magnet and the lid.

OBSERVATION: The magnetism will penetrate the thin cloth and paper as if it were not there. Substances containing iron will "short circuit" a magnetic circuit, although, if the magnet is strong, some of the magnetism will penetrate even a thick iron piece such as pliers, in sufficient strength to lift the lid underneath. The pliers here become magnetized while in contact with the permanent magnet.

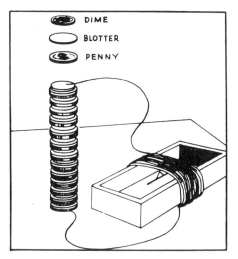

Money Power

MONEY POWER

NEEDED: Copper and silver coins, blotting paper, water, a way to measure the current.

EXPERIMENT 1: Stack the coins as shown: dime, blotter, penny, dime, blotter, penny, etc. As long as the blotters are wet, an electric current is produced.

The measuring device for the current shown here is a homemade galvanometer.

EXPERIMENT 2: It consists of a cardboard box wound with many turns of wire. Inside the wire coil a magnetized needle is suspended by a string so it is free to turn. Current in the coil can cause the needle to move. A pocket compass placed inside the coil works well.

Plain water may work in the battery, but it is much better if some sort of electrolyte is mixed with the water. A little salt, or vinegar, or lemon juice will serve.

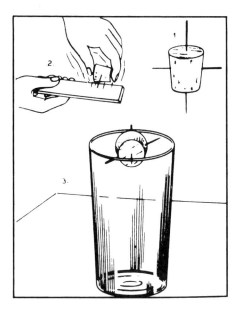

The Dip-Needle

THE DIP-NEEDLE

NEEDED: A long needle, a short needle, a cork, a drinking glass, a magnet.

EXPERIMENT: Stick both needles through the cork, and make them balance on the rim of the glass. Then magnetize the small needle by rubbing the eye of it on one end of the permanent magnet. When the needles and cork are again placed on the glass, with the small needle on a north-south direction, they will not balance level as before.

The needle will dip, to show that the earth's magnetic lines do not run parallel to the surface of the earth, but extend downward toward a point inside the earth. The amount of the dip will vary according to the locality of the earth on which the experiment is tried.

There is no dip at the magnetic equator, which is an irregular line running around the earth, in only a few places close to the geographical equator. Further, there are scattered regions far from the geographical equator where the dip is zero.

At the magnetic poles the dip is 90 degrees. Generally, the dip is small near the equator, and increases (in opposite senses) as you approach the North and South magnetic poles.

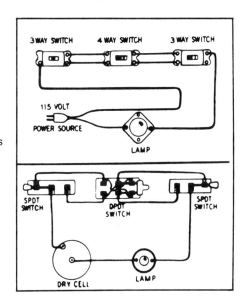

Tricky Switches

TRICKY SWITCHES

NEEDED: A double-pole, double-throw switch, two single-pole double-throw switches, a battery, a flashlight bulb in a socket, some wire.

EXPERIMENT: Make the connections as shown in the lower drawing, and by changing the positions of the switches, the lamp can be turned either on or off from either of the three switches. (Each switch must be closed one way or the other at all times.) The upper drawing shows how the connections are made in house wiring.

The switch in the center in both cases is a double-pole double-throw switch (abbreviated DPDT) and the switches on both sides are single-pole, double-throw (SPDT) switches.

No boy or girl should use the wiring method requiring line voltage from an electrical outlet—it can be dangerous. The same results are obtained with the lower wiring using a dry cell as the power source and a flashlight bulb as the light.

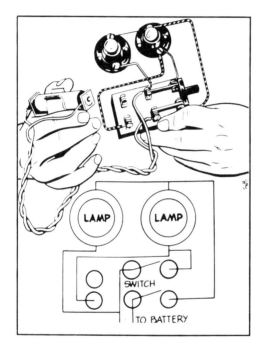

Dim the Lamps

DIM THE LAMPS

NEEDED: Two flashlight bulbs and a battery to match their voltage, sockets, a double pole, double-throw switch, wire for connections.

EXPERIMENT 1: Connect the lamps as shown. When the switch is closed in one position the lamps burn brightly because they both get the full voltage from the battery. They are connected in "parallel." If the switch is closed the other way the lamps are in "series" in which the same current must pass through one and then the other. Each lamp gets only half the total voltage and therefore is dimmer.

EXPERIMENT 2: It is assumed the two bulbs used in the

experiment on the preceding page are the same rating. Try using one bulb from a two-cell flashlight and one bulb from a three-cell light. If connected in series one will not burn as brightly as the other; more current can pass through the larger one than the small one can handle without full brightness. The smaller rated bulb will burn brighter than the other.

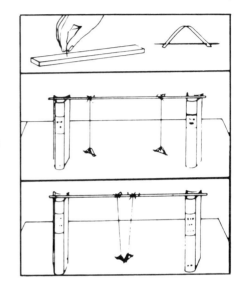

Compass Cut-Ups

COMPASS CUT-UPS

NEEDED: Two needles, a ruler, two books, a magnet, some thread.

EXPERIMENT: Magnetize two needles by rubbing the end of a magnet against them as shown in drawing 1, and suspend them on strings so that they will point North and South. Hang them on a ruler. Bring them together by sliding the strings along the ruler, and soon they will point toward each other, not North and South.

REASON: The magnets made by rubbing the needles produce magnetic fields close to them that are stronger than the magnetic field of the earth. The effect of their fields is noticed only when they are close together.

When they hang at a distance from each other, the earth's magnetism keeps them pointing North. The force of attraction of two magnet poles on each other is inversely proportional to the square of the distance between them. The strength of the fields does not vary.

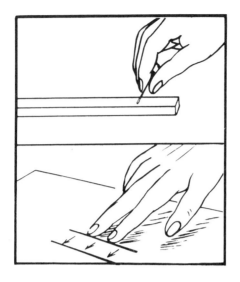

The Bashful Needle

THE BASHFUL NEEDLE

NEEDED: Two sewing needles, a magnet, a piece of glass or a smooth tabletop (not steel or iron).

EXPERIMENT: Magnetize the needles by stroking them with one pole of the magnet. Place one on the table, and approach it with the other. If the needles are attracted, turn one around. Then, as one approaches, the other rolls away.

REASON: If the needles approach so that the poles are at opposite ends, they will attract each other. But if they are brought toward each other so that a North pole of one approaches the North of the other, and the two South poles approach each other also, the needles repel each other. (The needle is best stroked flat against the magnet.) The rule is: like poles of a magnet repel; unlike poles attract.

THE BROKEN MAGNET

NEEDED: Two needles, two pairs of pliers, a magnet, a paper clip, a handkerchief.

EXPERIMENT: Magnetize a needle by rubbing it with one end of the permanent magnet. Note that either end of it will be attracted to the other needle.

Wrap the handkerchief around the magnetized needle, grasp it with the pliers, and break it near the middle. (The handkerchief is to prevent pieces of the needle from flying.) You now have two magnets, and their four ends will attract the other needle.

REASON: No matter how many times a magnet is broken, each piece becomes a magnet with a North and South pole. The new magnet will probably be of less strength because of the mechanical jar in breaking.

A Simple Electroscope

A SIMPLE ELECTROSCOPE

NEEDED: A jar, a metal wire, a rubber or plastic comb, some Christmas tree icicles.

EXPERIMENT: Bend the wire as shown, so it will hang in the jar, and bend an icicle so it hangs over the wire. Rub the comb vigorously on wool or fur, touch it to the wire, and the foil ends will swing apart.

REASON: The comb becomes charged by friction, and the foil and wire by induction as the comb approaches it. Since both ends of the foil carry the same charge, and since like charges repel, the foil ends push each other apart.

If humidity is high this instrument may not work. Warm it for a while in a warm oven to drive away moisture, then try it again. It should work readily in air-conditioned rooms or almost any heated room in winter.

Christmas tree icicles are made of aluminized Mylar film. The aluminized side must touch the wire. This is better than thin foil for this instrument. The electroscope may be charged by unrolling some types of cellophane tape near the wire rather than by rubbing a comb.

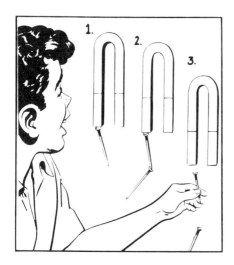

Magnetism by Induction

MAGNETISM BY INDUCTION

NEEDED: A magnet and two nails.

EXPERIMENT 1: Pick up one nail with the magnet as shown in drawing 1, and it will be found that the nail has become a magnet and will pick up another nail. This is magnetism by induction.

If the upper nail is pulled away from the magnet, as in drawing 3, the lower nail will fall off, because the upper nail will then have lost practically all of its magnetism.

Soft iron such as nails may be easily magnetized, but looses its magnetism just as easily.

EXPERIMENT 2: Try this with steel needles instead of nails. The bottom needle may not fall off as the top one is moved from the magnet. The top needle may become magnetized through contact with the permanent magnet so that it can hold the lower one. Steel does not lose its magnetism as easily as does softer iron nails.

RUG POWER

NEEDED: A wool rug, a rubber balloon, some bits of newspaper.

EXPERIMENT: Scatter the bits of paper on the floor. Rub the balloon briskly on the rug, hold it above the paper bits, and they will fly up to it.

Some of the paper bits will stick, while some may fly back down again.

REASON: Rubbing the balloon against the wool rug gives it a charge of static electricity which attracts the bits of paper. Some of them may then take up a charge from the balloon, and since it will be the same as that on the balloon, the paper and rubber will repel each other so that the paper flies away (like charges repel; unlike charges attract).

Static electricity experiments do not usually work in summer when the humidity is high. Cold weather outside, warm inside, is best.

Note that the electricity results from the motion the experimenter contributes to the experiment. If no energy is used, no energy is produced.

THE BASHFUL BALLOON

NEEDED: Two balloons, a wool skirt or cloth, a cold, dry day.

EXPERIMENT 1: Place one balloon on a table or floor. Rub the other on the wool until it is thoroughly charged with electricity. Hold it down so that it touches the other.

The second balloon will then be repelled by the first, and may be pushed around rapidly by it without being touched—as if too bashful to touch again.

REASON: When the balloon is rubbed against the wool, it takes on a minus charge. When it touches the second balloon, that too, becomes negatively charged. They repel each other according to the law that like charges repel and unlike charges attract.

EXPERIMENT 2: A balloon rubbed against wool until it is charged may be placed against the wall. It will remain stuck there; its charge attracts it to the wall, which remains uncharged because of its large area. A charged object is attracted to one uncharged or one carrying a smaller charge.

In these experiments the balloons may be charged usually by being rubbed against a rug on the floor.

EXPERIMENT 3: Rub a plastic or rubber comb through the hair. See if it will attract or repel the balloon.

MAGNETIZE THE SCISSORS

NEEDED: Scissors and a permanent magnet.

EXPERIMENT 1: To magnetize scissors, rub the points on one end of the magnet.

While magnetism is still not perfectly understood, it is believed that in magnetizing a piece of steel, the molecules are caused to align themselves in a more orderly fashion, parallel to one another.

Rubbing the points of the scissors on the other end of the

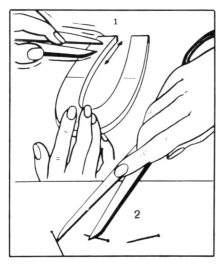

Magnetize the Scissors

magnet will de-magnetize them, and unless time is carefully checked, will remagnetize them with reverse polarity.

EXPERIMENT 2: If the points of the scissors are rubbed against the magnet as shown they both become either north or south seeking poles. Try rubbing one point against one pole of the magnet and the other against the other pole. The points should then have opposite polarity and should pick up pins better if held separated as in the lower drawing.

A MAGNET MYTH

NEEDED: A strong permanent magnet, two similar pieces of steel such as screwdrivers or needles.

THE MYTH: To make the screwdriver magnetic it must be stroked one direction on one pole of the strong magnet.

EXPERIMENT 1: Stroke one of the screwdrivers in the recommended way. It will become magnetized. Then rub the point of the other screwdriver back and forth on one pole of the strong magnet. It, too, will become magnetized, proving that the myth is only a myth. The steel need not be stroked one direction only.

A small magnet and sewing needles may be used to prove this, as shown in the drawing.

EXPERIMENT 2: Try rubbing one end of a needle on one pole of the magnet, rubbing the other end on the other pole. See whether the needle is magnetized more strongly this way.

EXPERIMENT 3: Try a long needle and a short one, and a thin one and a fat one. Which can be magnetized more strongly?

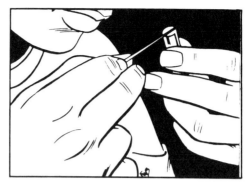

A Magnet Myth

THE OBEDIENT STRAW

NEEDED: Two soda straws, a jar, a fine thread, a rubber comb, a piece of woolen cloth or fur.

EXPERIMENT: Hang a piece of a straw in the jar as shown. Rub the comb briskly on the wool, then hold it near the jar. The straw inside the jar can be made to move about.

REASON: The comb becomes charged with an excess of electrons which are rubbed loose from the wool cloth. By induction, these repel electrons to the far end of the suspended straw. This makes the near end of the straw positive and attracted to the comb.

This and other static electricity experiments cannot be per-

The Obedient Straw

formed when there is much moisture in the air. They are performed best in a warm room in winter, or in an air-conditioned room where the humidity is kept low.

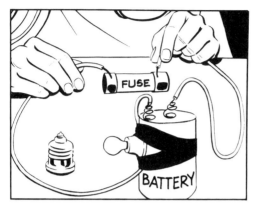

Make a Tester

MAKE A TESTER

NEEDED: A battery and bulb of the same voltage rating, wire, tape, soldering equipment.

EXPERIMENT: Connect the battery and bulb in series as shown in the drawing. If the battery terminals are not screw type the wires should be soldered to the terminals. The ends of the wires can be used to test fuses and some other electrical circuits.

REASON: If the wires are touched together the circuit is completed and the bulb lights. Touch the ends of the wires to the ends of a fuse, and if the fuse is good the bulb lights. If the fuse is burned out the bulb will not light.

Many other objects may be tested with this simple device, and it is safe because the battery voltage is low. Do not try to test anything that is plugged into an electrical outlet. The voltage there, probably 120 volts, is very dangerous.

THE CURIE POINT

NEEDED: Needles, a magnet, a candle or match, pliers.

EXPERIMENT: Magnetize a needle by rubbing one end of it on one end of the magnet. See if it will pick up other needles. Heat the needle over a flame, and again see whether it will pick up other needles.

REASON: If a magnet is heated to the "Curie point" the magnetism disappears. This effect was discovered by Pierre Curie in 1895.

The Curie Point

The many magnetic domains in the needle are initially arranged at random, and do not make the needle a magnet to any appreciable degree. When the needle is rubbed on the magnet some of the domains line up so their magnetism points in the same direction. Their magnetic domains add together rather than canceling one another out.

Heat makes the domains return to the random state. The Curie point, which varies with different alloys, is the point at which the magnet loses its magnetism as the domains return to the random state.

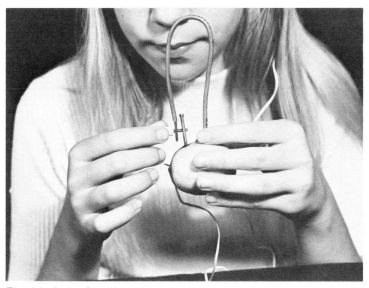

Electricity from a Lemon

ELECTRICITY FROM A LEMON

NEEDED: A lemon, a common nail, copper wire, an earphone from a transistor radio.

EXPERIMENT 1: Stick the nail and the end of the copper wire into the lemon. Bend the wire so the other scraped end comes close to the nail. Touch the earphone plug to the nail and wire at the same time, and static sounds will be heard in the earphone.

EXPERIMENT 2: Try this with a potato and other vegetables and fruits.

REASON: The lemon juice, which is the electrolyte for this small cell, reacts at a different rate on the iron and copper, causing a small potential or voltage difference to exist between them.

When the earphone plug makes contact with the metals a current flows in the earphone wires, causing the static sounds. A sound is heard in the earphone when contact is made and again when it is broken. A different sound is heard if the plug is made to slide along the wire and nail.

Chapter 9

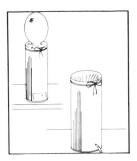

Air, Air Pressure & Gases

A SODA STRAW ATOMIZER

NEEDED: A glass of water and a soda straw.

EXPERIMENT: Cut the straws in about half. Place one in the water, the other in the mouth, and blow as shown in the drawing. Water will be seen to rise in the straw. If we blow harder, the water

A Soda Straw Atomizer

will rise to the top of the straw and will be sent in fine droplets across the table.

REASON: Moving air exerts less lateral pressure than the still or less turbulent air around it. The moving air at the upper end of the straw reduces the pressure in the vertical straw so that atmospheric pressure on the surface of the water in the glass can force some up into the straw.

"Atomizer" is an incorrect word. The water spray is in the form of small drops, not "atoms."

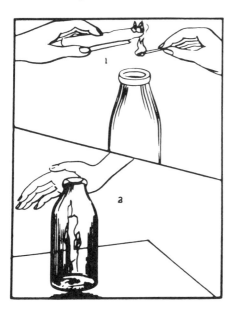

A Milk Bottle Octopus

A MILK BOTTLE OCTOPUS

NEEDED: A wet milk bottle, a piece of tissue paper, a match.

EXPERIMENT 1: Twist the paper into a loose rope form and light it with a match. Drop the lighted paper into the bottle, and place the wet hand over the opening. As the flame goes out, the strong suction felt may give some idea of how the octopus' suction cups can draw blood through the skin. The bottle can be lifted as shown in the lower drawing. But don't hold it too long.

REASON: The flame heats the air in the bottle, making it expand so that some of it escapes. When the flame goes out, the cool bottle cools the air, making it contract. This reduces the pressure, so that the pressure in, above and around the hand forces the skin downward into the bottle neck.

112

Don't use too much paper; the flame might burn the hand.

EXPERIMENT 2: Try this experiment without fire, as shown in the photo. Fill the bottle with boiling-hot water, invert the bottle over the sink so the water gurgles out, then put the wet hand over the mouth of the bottle. Condensing steam and cooling vapor reduce the pressure in this case.

Milk bottles are scarce most places. Other bottles may be used; the author uses a bottle that contained white Karo syrup. Try the Karo bottle in other experiments where a milk bottle is required.

A BALLOON TRICK

NEEDED: A jug or bottle, a balloon, a refrigerator.

EXPERIMENT 1: Cool the jug in the refrigerator an hour. Take it out, put the collapsed balloon tightly over the mouth of the jug, and let it stand. As it warms, the balloon will be blown up.

EXPERIMENT 2: Another way to show the same principle: place the balloon over the mouth of the jug at room temperature, then heat the jug under the hot water faucet.

REASON: As the air in the jug becomes warmer, it expands, and some of it is forced out into the balloon. (Look up Charles' Law.)

Balloon in a Jug

BALLOON IN A JUG

NEEDED: A jug, a balloon, hot water, a hard-boiled egg.

EXPERIMENT 1: Boil the water, and while boiling hot, pour some into the jug. Shake it around, and pour it out. Place the balloon over the mouth of the jug—quickly. Watch.

REASON: As the steam and hot vapor in the jug cool, they require less space. Pressure in the jug is reduced, and pressure of the outside atmosphere inflates the balloon—pushing it inside the jug.

EXPERIMENT 2: This is the same principle used to put a hard-boiled egg into a glass milk bottle. Peel the egg first, place it over the mouth of the bottle, and watch it plop in. The egg must be placed over the bottle quickly after the hot water is poured from the bottle.

EXPERIMENT 3: To get the egg out, blow into the bottle while holding it inverted. To get the balloon out of the jug, take it loose from the mouth of the jug and place a pencil down beside it to allow air to flow into the jug around the balloon. The balloon deflates quickly and can be pulled out.

Lung Power

LUNG POWER

NEEDED: Book, rubber balloon, a table.

EXPERIMENT 1: Place the balloon under the book, as shown, and as air is blown into the balloon, the book will rise.

REASON: Air pressure from the lungs is sufficient not only to lift the weight of the book, but also to stretch the rubber of which the balloon is made.

A much heavier weight can be used if a hot water bottle is substituted for the balloon.

EXPERIMENT 2: The author, in one of his school demonstrations in science, places a beach ball between hinged boards and lets a pupil stand or kneel on it. Another pupil can lift the person by blowing breath into the ball.

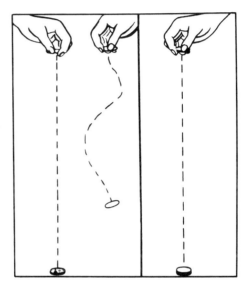

A Paper and Coin Trick

A PAPER AND COIN TRICK

NEEDED: A coin (a half dollar or quarter will do) and a piece of paper.

EXPERIMENT: Cut the paper so that it is slightly smaller than the coin. Leave the paper flat so that it will lie flat on the coin. Drop both paper and coin separately as shown in the drawing at left, and the paper flutters down more slowly than the coin falls. Drop both together and they fall together, as shown at right.

REASON: The heavy coin, as it falls, takes some air along with it, and the paper rides in this "captive" envelope of air. Sometimes the paper gets separated from the coin during the fall, and then it flutters down slowly because it leaves the air envelope surrounding the coin.

In a vacuum, both would fall at the same rate even though they were dropped separately.

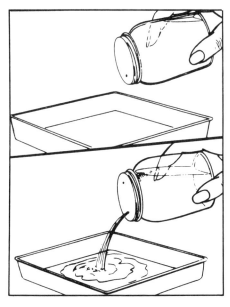

A Bottle Trick

A BOTTLE TRICK

NEEDED: A jar with a tight metal lid, a nail, some water.

EXPERIMENT: Make a nail hole in the lid as shown in the upper drawing. Put water into the jar, and hold it so that the water should run out. It does not.

Make another hole in the lid, as in the lower drawing, and the water pours out.

REASON: A little water may pour out at first, but not much, because the pressure of the atmosphere and surface tension hold it in. If a second hole is made, so that air may get in to make the pressure on the water at the lower hole in the jar exceed that outside the jar, the water can pour out.

And why doesn't the air go in and the water out of the one hole at the same time? The surface tension of the water acts as a film to close the one little hole.

THE CHICKEN FOUNTAIN

NEEDED: A milk bottle, some water, a drinking glass.

EXPERIMENT: Invert the bottle of water in the glass, and a little water will come out. Raise the bottle, and the water in the glass will rise just as far as the mouth of the bottle is raised.

REASON: There is a force on the surface of the water in the glass due to atmospheric pressure. Unless there can be a greater

The Chicken Fountain

force downward in the neck of the bottle, the water cannot run out. This force cannot exist unless air can get into the bottle, and air cannot easily get in until the bottle is raised above the surface of the water in the glass.

The illustration shows the practical use of this principle in the poultry drinking fountain, where water runs from the jar only as the chickens drink from the open part of the container. As the chickens drink from the lower reservoir, air is admitted to the jar, allowing an equal volume of water to run out.

The pressure of air at sea level is equal to a column of water about 34 feet high.

WEIGHT OF AIR

NEEDED: A rubber balloon, a yardstick, a wire, some string.

EXPERIMENT 1: Hang the wire, balloon, and the stick as shown. Move the wire until the stick is balanced. Puncture the balloon. Note that the wire is now heavier than the punctured balloon.

REASON: The air in the balloon was compressed by the balloon; therefore it was more dense and heavier than an equal volume of air at regular atmospheric pressure.

EXPERIMENT 2: Two balloons may be used instead of a

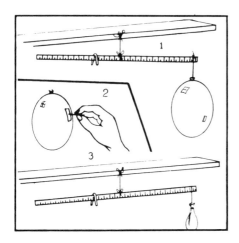

Weight of Air

balloon and a weight. This experiment is often presented as a demonstration that air has weight. That is not true. It shows only that air compressed by the tight rubber has greater weight than an equal volume of air at regular atmospheric pressure.

INSIDE-OUT BALLOON

NEEDED: A tin can with the end cut out smoothly, a piece of rubber balloon, a nail, some water, some string, a hammer.

EXPERIMENT: Punch a hole in the side of the can near the

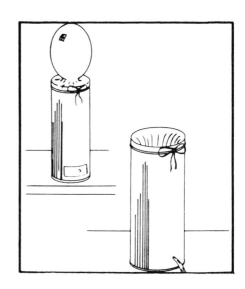

Inside-Out Balloon

bottom, cover it with plastic tape, and fill the can with water. Stretch the rubber over the end of the can and tie it tightly.

Set the can upright, take off the tape, and as the water comes out of the can, the rubber will be pushed inward by the invisible molecules of air.

REASON: As the water leaves the can, the pressure of the air forces the rubber inward. If there is a hole in the balloon, air will leak in to fill the space left by the receding water, and the balloon will remain straight because the air pressure will then be the same on both sides of it.

If the balloon can be tied on the can as shown in the upper drawing, it will turn inside out as the water leaves the can.

The Hanging Glass

THE HANGING GLASS

NEEDED: A glass of water and a flat rubber sink stopper.

EXPERIMENT 1: Wet the stopper, press it down on the glass, and both the stopper and glass may be lifted by the ring.

REASON: When the ring is lifted, the rubber pulls upward slightly, decreasing the pressure of the air in the glass. Atmospheric pressure then presses the rubber to the glass so firmly that the glass of water can be lifted before the rubber will pull loose.

Wetting the rubber before the experiment makes a better

air-tight seal between the rubber and the rim of the glass. Surface tension of the water helps in the lifting operation.

EXPERIMENT 2: If a larger demonstration of this principle is wanted, a rubber suction cup known as a "plumber's helper" may be used. These come with wooden handles. Moisten one, push it down on a smooth surface, and it may be impossible for the normal person to pull it loose. If the area touching the surface is 20 square inches, the pressure of the atmosphere on it can be 280 pounds or more.

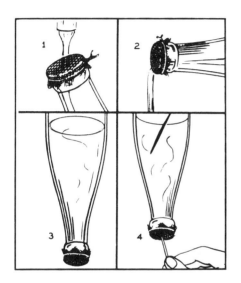

A Mystery Bottle

A MYSTERY BOTTLE

NEEDED: A soft drink bottle, screen wire, water, toothpicks.

EXPERIMENT: Cover the mouth of the bottle with the screen wire. Fill it with water, as shown in 1. The water pours in freely, and pours out just as readily if the bottle is held as in 2.

But turn the bottle upside down, as in 3, and the water does not pour out. Toothpicks may be inserted through the mesh of the screen; they will float to the top of the water and still the water does not pour out.

REASON: Surface tension of the water is like a thinly stretched rubber sheet. The screen wire increases the surface tension so that the film does not break, and the water does not run out.

The atmosphere exerts small upward force on the lower surface of the film because of the slight decrease in pressure of the air in the bottle.

ELASTIC SOAP BUBBLES

NEEDED: A thread spool, a candle, bubble solution.

EXPERIMENT: Blow a bubble on the end of the spool, hold the spool as shown, and as the bubble gets smaller, the air from it will blow the flame.

REASON: The soap and water film that makes up the bubble is elastic, much as a rubber balloon would be. This is another experiment in surface tension, demonstrating that force due to surface tension draws the surface of a liquid into a spherical shape.

Elastic Soap Bubbles

MAKE A DRAFT

NEEDED: A cardboard box, cellophane, cellophane tape, two candles, a match.

EXPERIMENT: Cut a hole in the side of the box and cover it with cellophane, holding the cellophane with the tape. Make a hole the size of a half dollar in the end of the box. (Make another hole the same size in the other side of the box as shown.) Light both candles. Place one on the table, and place the box over it so that the hole is over the flame. Hold the other candle flame near the other hole, and a draft will be seen which will suck the second flame into the box.

OBSERVATION: As air is heated it expands, becomes lighter, and rises as heavier, cooler air is drawn in to replace it. This is very much the way winds are formed in the air over the earth.

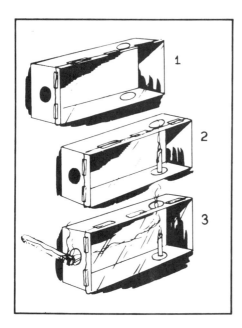

Make a Draft

THE RISING PAPER

NEEDED: A strip of paper.

EXPERIMENT: Hold the paper at the mouth, blow over it, and it will rise to the horizontal position.

REASON: Bernoulli learned that a moving air current has less side pressure than the still or slower-moving air beside it. When air is blown above the paper, it has less pressure than the air below. The paper is then pushed up by the greater air pressure below.

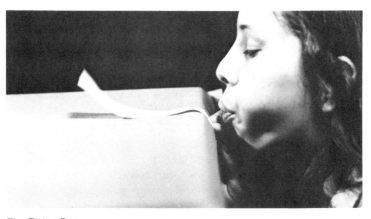

The Rising Paper

Why No Blow-Up?

WHY NO BLOW-UP?

NEEDED: A sheet of paper and two books.

EXPERIMENT: Place the paper on the books, blow straight under it, and it will bend downward, not upward as expected.

REASON: Bernoulli discovered that air in motion exerts less lateral or side pressure than air at rest or moving more slowly. When air is blown under the paper, it therefore exerts less pressure than the still air above it. The still air then pushes the paper down. This is the principle by which airplanes fly.

THE DANCING BALLOON

NEEDED: A rubber balloon and a warm air register in the floor.

EXPERIMENT: Place the balloon in the stream of rising air, and it will remain there. If the air stream is strong, the balloon will dance up and down in the air.

REASON: Bernoulli discovered that air in motion exerts less side pressure than still or more slowly-moving air. Therefore, if the balloon moves to one side of the warm air stream, the greater pressure from the slowly-moving air surrounding the moving stream pushes the balloon back.

There is also an irregular up and down dance representing a contest between the downward pull of gravity and the upward

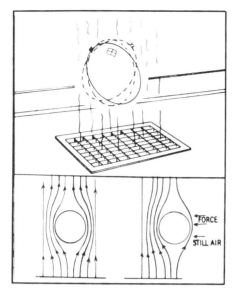

The Dancing Balloon

movements of the irregular balloon due to the ascending air currents.

THE MISCHIEVOUS BALL

NEEDED: A funnel, a table tennis ball, a hose-type vacuum cleaner.

EXPERIMENT: Place the hose in the cleaner so that the air blows out. Place the funnel and ball as shown in Drawing A, and the ball will float above it. Place the ball down into the funnel, and it sticks, as in B. Hold the funnel and ball downward as in C, and the ball will be sucked into the funnel, although the air is blowing out.

REASON: The Bernoulli effect is that the lateral pressure of a stream of fluid (air is a fluid) is decreased as the rate of flow is increased. In A, the moving air exerts less pressure than the still air around it, and so the still air pushes the ball back on the moving stream when it tends to move out of it.

B. The moving air around the ball exerts less pressure than the still air pushing down on the top of the ball.

C. The same applies here, except the still air pushes on the bottom of the ball.

A HEAT MOTOR

NEEDED: A paper square, scissors, spool, pencil, needle, thimble, a match.

124

The Mischievous Ball

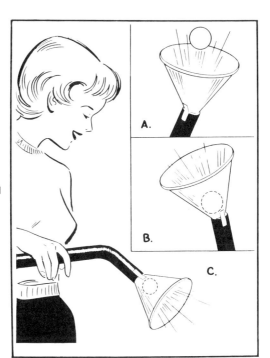

A.

B.

C.

A Heat Motor

EXPERIMENT: Cut the paper into a spiral as shown in the upper right drawing. Make a pinhole in the center. Mount it on the pencil as shown, hold a lighted match below it, and the spiral will turn.

REASON: The match warms the air above it. The lighter warm air rises and hits the bottom surface of the spiral, causing it to turn due to the unbalance of forces.

Blowing on the spiral will also make it turn.

THE BLOW-BOTTLE

NEEDED: A bottle and a lighted candle.

EXPERIMENT: Place the thumb over the end of the bottle. Put both thumb and bottle into the mouth. Release the thumb slightly while blowing hard into the bottle, then cover immediately with the thumb.

Bring the mouth of the bottle, with your finger still covering it, close to a lighted candle. Then remove your finger.

RESULT: The air inside the bottle will rush out with enough force to make the candle flicker as though you had blown on it.

While the lungs are not a very good air compressor, they are strong enough to compress enough air for this experiment. While the air in an automobile tire is perhaps 30 pounds per square inch, above atmospheric pressure, the human lungs can offer pressure of only about 1.5 pounds per square inch above atmospheric pressure. (The term "gauge pressure" may be used rather than "above atmospheric pressure.")

THE PARACHUTE

NEEDED: A handkerchief, some string, a weight.

EXPERIMENT: Make the parachute as shown, throw it into the air so that the weight *pushes* the cloth upward. As it comes down, the cloth spreads and the parachute descends slowly.

REASON: As the chute goes up, it is in a compact mass, so that air resistance is slight. When the cloth opens out it must move a larger amount of air out of the way as it comes down, and this means more air resistance and slower motion.

A man coming down with an open parachute hits the ground with about the same speed as though he jumped from a 10-foot height.

THE ELEVATOR CARD

NEEDED: A thread spool, some thread, some cardboard squares, glue.

The Parachute

EXPERIMENT 1: Paste one cardboard square to the spool as shown in the drawing at center right. There must be a hole in the center of the card to match the hole in the spool. Hang the other card to the first with thread passed through holes at the corners.

Blow downward through the spool, and the lower card will rise up against the upper card.

REASON: The lower drawing explains; air in motion between

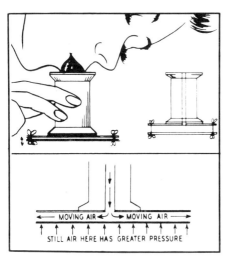

The Elevator Card

MOVING AIR ← → MOVING AIR
STILL AIR HERE HAS GREATER PRESSURE

the cards has less lateral pressure than the still air that pushes up against the bottom of the lower card. This is the Bernoulli principle,

EXPERIMENT 2: The author presented the simpler version of this in one of his earlier books. Take a 2-inch square of light cardboard, push a straight pin into the center of it, place it on a spool so the pin keeps the card from sliding off, then try to blow the card off by blowing up through the hole in the spool.

Chapter 10

Heat

EXPLOSIONS ON THE KITCHEN STOVE

NEEDED: Popcorn, a covered skillet or popper, a stove.

EXPERIMENT: Pop the corn.

OBSERVATION: Most substances when heated give off their moisture slowly. The water in a grain of popcorn, however, is enclosed in an air-tight sheath which does not explode until a rather powerful steam pressure is built up inside it.

As the sheath bursts, most of the cells inside it explode, too, from the steam pressure inside them. The expanding cells form the delicious white mass.

The pressure that builds up inside the popcorn before it pops has been estimated at from 15 to 100 pounds per square inch.

THE OBLIGING BALLOON

NEEDED: A small rubber balloon, two water glasses, hot and cold water.

EXPERIMENT: Blow up the balloon. Heat both glasses by pouring hot water into them. Place them quickly on either side of the balloon, hold them securely, then cool the glasses under the cold water faucet.

The sides of the balloon will be sucked up into the glasses, and will hold so tightly that one glass may be lifted with the other as shown in the drawing.

REASON: As the glasses are cooled under the cold water, the

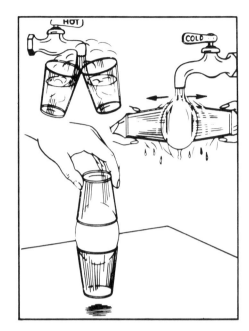

The Obliging Balloon

air in them cools and contracts. The reduced pressure allows the pressure of the atmosphere to force parts of the balloon tightly into the mouths of the glasses.

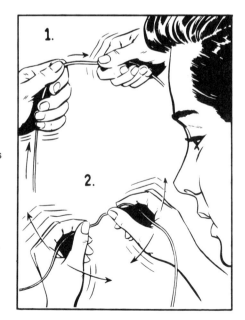

Two Frictions

TWO FRICTIONS

NEEDED: A piece of bare wire. A wire coat hanger will do.

EXPERIMENT 1: Pull the straight wire through the hand. Heat will be felt.

EXPERIMENT 2: Bend the wire back and forth quickly several times. It will get warm.

REASON: The answer is friction in both cases. Friction between the wire and hand generates heat. In the second experiment the friction is between molecules of the wire as they move over each other.

In both instances mechanical energy is converted into heat energy by means of friction.

MYSTERY ICE

NEEDED: Soft drinks in the freezer.

EXPERIMENT 1: Take out a bottle of the drink just before it begins to freeze. Ice will form in it when the cap is removed.

EXPERIMENT 2: Don't let any liquid freeze solid in a glass container in the home freezer. Water expands when it freezes and will break the strongest glass bottle. To show this, put water into a bottle, wrap the bottle several times with a cloth to protect from broken glass, and let it freeze.

REASON: Carbon dioxide gas dissolved in the drink lowers its freezing point. When the bottle cap is removed, some of the carbon dioxide will escape, and this raises the freezing point so that it can freeze at a higher temperature.

AN OLD FREEZING QUESTION

NEEDED: Boiled water, boiled and aerated water, tap water, salt water; four similar glasses.

EXPERIMENT: Boil some of the water for two minutes. Fill one of the glasses with it. Put some of the boiled water into a jar, tighten the lid, and shake it vigorously to get some air dissolved in it. (This should be done after the water has cooled.) Fill a glass with this aerated water, another with plain tap water, and another with salt water.

After all the glasses of water have been allowed to reach the same room temperature, place them into a freezer.

OBSERVATION: The boiled water may freeze first, the aerated water second, the plain water third, and the salt water last. The reason is that most substances added to water lower its freez-

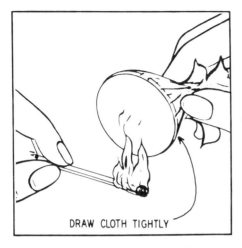

The Fireproof Handkerchief

DRAW CLOTH TIGHTLY

ing point. This includes air. Boiling releases most of the air that is dissolved in water.

THE FIREPROOF HANDKERCHIEF

NEEDED: A half dollar, a handkerchief, a match or cigarette.

EXPERIMENT: Wrap the coin in a single thickness of the cloth, and draw the cloth tightly around the coin. A match flame held to the tight part of the cloth for a short time will not burn it. The lighted end of a cigarette may be placed against the cloth without burning a hole in it.

REASON: While the cloth does not transmit heat very well, it lets enough go through into the metal coin to keep the temperature of the cloth below its burning point. The metal conducts the heat away from the cloth which is in contact with the coin.

(Suggestion: do not use a valuable handkerchief for this experiment—it *might* get scorched.)

SQUEEZE BOTTLE

NEEDED: An empty soft drink bottle, a dime, some water.

EXPERIMENT: Place the dime over the mouth of the bottle, and drop a little water around the edge to seal it. Grasp the bottle with both hands, squeeze it, and the dime will dance up and down.

REASON: Squeezing the bottle probably has no noticeable effect on the dime at all, but the warmth of the hands causes the air inside the bottle to expand after the squeezing has been going on for several seconds (Charles' law). The escaping air causes the dime to move up and down.

132

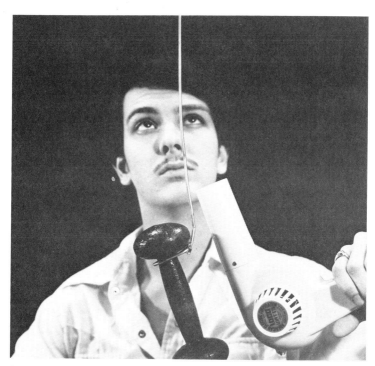
The Hot Weight

Charles' law teaches that the volume of any gas is directly proportional to its absolute temperature.

THE HOT WEIGHT

NEEDED: Wire, a weight, ice cubes, a hair dryer.

EXPERIMENT 1: Hang the weight on the wire so it barely swings above a bare table or floor (no carpet). Heat the wire with the hair dryer, and it will expand so that the weight drags the floor or table.

EXPERIMENT 2: The wire may be cooled enough by running the ice cubes over it to make it contract so the weight swings freely again.

REASON: Most substances expand when warmed and contract when cooled. Iron, copper, or aluminum wire will demonstrate this in this experiment.

HOT OR COLD

NEEDED: Breath on a cool day.

Hot or Cold

EXPERIMENT: Hold the hand in front of the open mouth and blow. The air is warm. Blow through pursed lips, and the air is cool.

REASON: Air coming from the open mouth comes at little less than the temperature of the inside of the body, which is warm. But when air is forcibly blown in a thin steam through the lips it feels cooler for two reasons.

The air is compressed slightly and cools when it expands outside the mouth. This is one of the laws of science: a gas cools as it expands.

The air stream coming rather fast from the lips gathers surrounding air into the stream, and that surrounding air is cool.

FREEZING

NEEDED: Olive oil.

EXPERIMENT: Place the bottle of oil in the refrigerator (not the freezing compartment) and let it freeze. Note that when it is removed and begins to thaw, the frozen oil will be on the bottom, and not floating on the top as in frozen water.

REASON: Most liquids, when they freeze, become solid from bottom to top, because they contract on freezing. Water expands when it freezes, so is lighter, and the solid ice floats on top. This is

Freezing

fortunate for living things in water. If water did not freeze in this way, our lakes and rivers could become solid ice.

CRAZY WATER

NEEDED: Hot and cold water spigots.
EXPERIMENT: Turn on a little water from both spigots. The

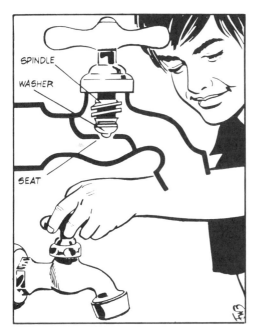

SPINDLE

WASHER

SEAT

Crazy Water

135

cold water will continue to run, but chances are the hot water will turn itself off after a few seconds.

REASON: The water flow is controlled by a metal-and-rubber plunger that moves in or out with a screw as the handle is turned. Opening the spigot (more correctly called a faucet) moves the plunger out so water can flow past it.

Hot water causes the plunger to expand so that it closes the water opening. The cold water does not cause the cold water plunger to expand.

THE WARM SLEEVE

NEEDED: Two sleeves of different weaves, sunlight.

EXPERIMENT 1: Hold the arms in the sun. The arm covered with tightly woven material will get warm more quickly than the other.

EXPERIMENT 2: Try this with a dark and a light sleeve in the same weave. The dark will absorb light, changing it to heat, much more than the light color.

REASON: The loosely woven material has many airspaces between the threads, and air is an insulator against heat. The solidly woven material conducts heat more readily.

It is important to use material of the same color for this comparison, since some colors absorb more heat than others and would conduct it on to the arms. A white sweater with tight sleeves, and a cotton shirt sleeve drawn tightly against the arms were used in the author's test of this experiment.

STARCHY BUBBLES

NEEDED: A teaspoon of starch in half a glass of water, a

The Warm Sleeve

Starchy Bubbles

saucepan with handle, an electric stove.

EXPERIMENT: Let the starch and water mixture come to a boil. Press down on the pan, and the boiling is more rapid as shown by the increase of bubbles. The starch makes the bubbles more visible.

REASON: Pressing down on the pan makes better contact of the pan with the heating element. The elements are seldom perfectly flat, and when the pan is pressed, more of the heating surface touches the pan, for two reasons. Pressing tends to flatten the element, and thus gives more contact. Also, pressing tends to flatten the pan against the element somewhat, although this cannot be seen with the eye.

This second type of flattening takes place when any two substances are pressed together. Even a heavy railroad rail is flattened somewhat when the wheel rolls over it. The metals usually return to their original shape when the pressure is removed.

FAST COTTON

NEEDED: Strips of polyester cloth, strips of cotton cloth, water.

EXPERIMENT 1: Dip the cloth pieces into water. The cotton soaks up water immediately; the polyester does not.

EXPERIMENT 2: Get a cotton cloth and a permanently pressed cloth (shirttails of good shirts will do—this will not damage the shirts) and dip them in water. The plain cotton will soak up water more quickly.

REASON: Why does a dress or shirt made of polyester material feel hotter on a summer day than one made of cotton? The

Fast Cotton

structure of the threads in the different materials is quite different. The cotton cloth fibers have spaces that allow water to flow up between them by capillary attraction and heat from the body to flow out from the body through the fibers. The polyester fibers are more solid. The wrinkle-proofing substances in the permanently pressed cloth will slow or prevent the easy flow of water through the fibers by capillary attraction.

Chapter 11

Light

A SKIN-DIVER'S MYSTERY

NEEDED: A glass of water and a pencil.

EXPERIMENT: Place the pencil in the glass and look up at it as shown in the drawing. The pencil—and other objects—cannot be

A Skin-Diver's Mystery

seen when they are above the surface of the water.

REASON: For light striking the under surface of water at an angle of about 48.6 degrees or more, that surface acts as a mirror, reflecting the rays back. The exact angle depends on the temperature of the water.

This is why a skin diver who looks up can see out of the water only at an angle less than 48.6 degrees from the vertical.

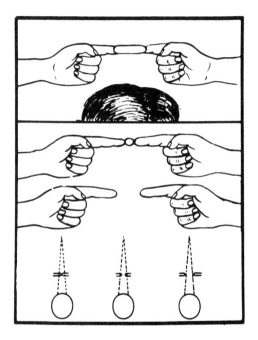

A Sausage from the Fingers

A SAUSAGE FROM THE FINGERS

NEEDED: Just two fingers.

EXPERIMENT: Hold the fingers together in front of the face, at arm's length, and look at a distant object, as in the diagram. The fingers seem to form a sausage as in the top drawing. Draw the fingers apart somewhat, as in second diagram, and the sausage becomes a ball. Draw them farther apart, as in third diagram, and the ball disappears.

REASON: The dotted lines in the diagrams show the path of light from the distant object (where the lines converge) to the two eyes. The circles represent the head. This is an optical illusion.

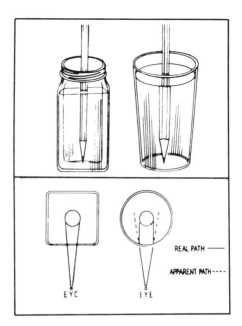

Glasses That Magnify

REAL PATH ————
APPARENT PATH ----

EYE   EYE

GLASSES THAT MAGNIFY

NEEDED: A round glass container, a square glass, water, a pencil.

EXPERIMENT: Place the pencil in the square glass of water, and it looks the same size. Place it in the round glass of water, and it looks larger where it is under the water.

REASON: The curve of the glass acts as a magnifying glass. The lower drawings show the real and apparent paths of light coming from the pencil to the eye through the glasses.

There is refraction or bending of the light at both surfaces of the glass, changing the direction of the light ray. The light appears to come to the eye along the dashed line, as though the pencil were larger.

BIRD IN A CAGE

NEEDED: Twine, a piece of cardboard, ink or crayon.

EXPERIMENT: Draw a bird on one side of the card, and a cage on the other. Attach the strings as shown, twist them, and pull on the ends to make the card spin around rapidly. The bird will be seen in the cage.

REASON: The eye continues to see an object after it has disappeared. This lasts only a fraction of a second, and is called

141

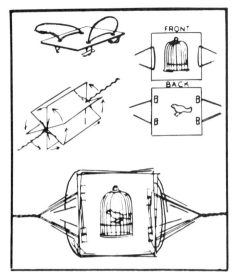

Bird in a Cage

"persistence of vision." Therefore, the eye continues to see both pictures on the card as it turns.

It is persistence of vision that allows us to see movies and television as constant pictures that move, instead of what they are: a series of flickering pictures, alternating with blank screens.

Why the Dark Spots?

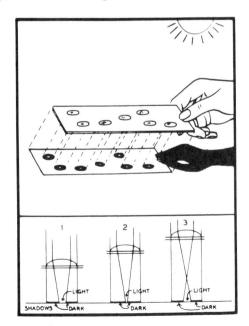

WHY THE DARK SPOTS?

NEEDED: A piece of glass, some drops of water, a light surface, some sunshine.

EXPERIMENT: Place drops of water on the glass, and hold it above a sheet of white paper or other light surface. The water is clear, yet there will be dark spots on the paper below the water drops, many with an extra light spot in the center. Move the glass up and down.

REASON: The sunlight striking the glass shines through evenly. But the light striking the drops is refracted so that it is concentrated in the center and surrounded by dark circles. The diagram in the lower part of the drawing shows this refraction.

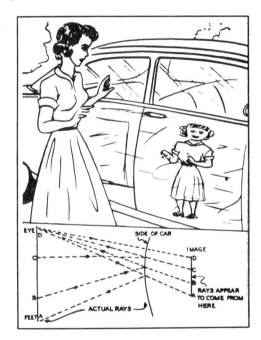

The Fat Reflection

THE FAT REFLECTION

NEEDED: A shiny automobile door.

EXPERIMENT: Look at a reflection of a person in the door of the car. The person will look fat.

REASON: The diagram shows the paths of the light rays which travel from various parts of the body to the car door and back to the eye of the viewer.

Rays from the curved door are spread outward, up and down,

143

but not right and left, because the door surface is not spherical. The width of the reflection is about the same as the person, but the height is much less, making it out of proportion.

Two Mirror Tricks

TWO MIRROR TRICKS

NEEDED: Two mirrors, some cellophane tape, a short pencil.

EXPERIMENT 1: Tape the mirrors together at right angles to each other. Stand them up, and place the pencil between them. How many pencils do you see?

EXPERIMENT 2: Look into the mirrors, and the face will be seen, but not as in an ordinary mirror.

REASON: Light leaving the pencil goes in all directions. Some passes directly to the eye, some passes to a mirror then to the eye, and some may be reflected from both mirrors before reaching the eye.

The face is reflected from both mirrors before reaching the eye. This gives an effect opposite to that from a single mirror. A finger touching the left side of the face seems to touch the left side of the mirrored face.

THE CORNER REFLECTOR

NEEDED: Two dime store mirrors, a flashlight.

EXPERIMENT: Arrange the mirrors as shown, so the angle is 90 degrees between them. They may be held in place with tape. Then, when the light is shined at the angle where the mirrors meet, or anywhere within the angle created by the mirrors, the light reflects back to its source. If the light is held below the arrangement, another mirror placed on top of the two is necessary. If the light is held above, another mirror placed below the two is necessary.

An array of such mirror combinations was left on the moon so a laser beam could be reflected back to earth from it.

REASON: A more technical explanation is: the corner reflector has the property—from elementary geometrical optics—that a ray of light impinging on it is reflected back parallel to the incident ray.

Small mirrors are good for this experiment.

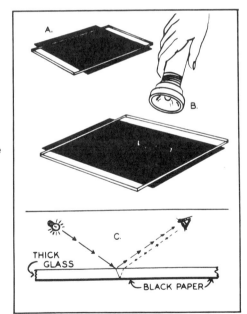

The Double Image

THE DOUBLE IMAGE

NEEDED: A piece of thick, flat glass, black paper, flashlight.
EXPERIMENT: Place the black paper behind the sheet of

glass and you have a mirror. But shine the flashlight on your new mirror as shown in Drawing B. You will see that two images are reflected from the mirror.

REASON: Both the top and bottom surfaces of the glass act as mirrors. The light rays coming from the flashlight are reflected and refracted as shown in Drawing C.

Light going from a transparent substance into another transparent substance of a different density is bent or refracted. This may be seen in Drawing C.

A Rock Reflector

A ROCK REFLECTOR

NEEDED: A small stone or other rough object, clear nail polish.

EXPERIMENT: Paint half the stone with the polish, or simply wet half of the stone with plain water. Note how much more glossy the painted or wet half appears.

REASON: Light striking the rough stone is reflected in an irregular fashion which is called diffused reflection. This is how we see most objects. The polish or water on the stone makes the surface appear as many small mirrors reflecting the light in a more regular pattern as from many small smooth surfaces.

AN ILLUSION

NEEDED: Smoked glass or dark sunglasses, a weight on a string (the author used a bar of soap wrapped in foil).

EXPERIMENT: Have someone cover one eye with the dark glass, then, leaving both eyes open, watch the weight. Swing the weight in a straight arc perpendicular to the other person.

He is likely to see the weight swinging in a circle. Have him

An Illusion

switch the glass to the other eye, and the weight will seem to go around the other direction.

COMMENT: The author has not found agreement among scientists as to the cause of this illusion. He invites opinions. Note that not everyone will observe the circular motion of the weight.

This works better if the people performing it are several feet apart. The pendulum must swing at right angles to the line of sight, not to and from the viewer.

THE TYNDALL EFFECT

NEEDED: Two coffee cans, tape, an electric lamp and socket, a glass container such as a fish bowl, water, milk, a dark room, cardboard.

EXPERIMENT: Mount the lamp in the bottom of one can. Make a round hole in the bottom of the other can (a large nail can punch the hole) then tape the cans together as shown. Place the bowl of water where the light beam will shine through it, and add

The Tyndall Effect

one or two drops of milk. A disc of cardboard with a hole in its center goes in the end of one of the cans.

The beam of light seen in the water is called a Tyndall cone, after John Tyndall, a British scientist, although the effect was first seen by Faraday in 1857.

REASON: The light is reflected or "scattered" by the tiny colloidal particles from the milk, and is polarized. Various sizes of particles give various colors to the cone, which is only slightly cone shaped after all, but is extremely interesting to a technical man. Tyndall made the first exhaustive study of it. If the particles are smaller in diameter than one-twentieth the wavelength of light, blue is the predominant color.

The definition of Tyndall Effect is: visible scattering of light along the path of a beam of light as it passes through a system containing discontinuities. (McGraw-Hill *Encyclopedia of Science and Technology.*)

Without the cardboard disc between the cans the light does not come out in a narrow beam but reflects many ways from the insides of the cans. Spraying the insides of the cans with dull black will eliminate the need for the cardboard disc.

Mach Bands

MACH BANDS

NEEDED: A shaded lamp, white card, another card, any color, a darkened room.

EXPERIMENT: Hold the card as shown so the light falls on half of the white card. Notice that where the shadow begins there is a light streak next to the bright reflection from the card, and a dark streak on the dark shadow side of the card.

REASON: Of course there are really no such lines; this is an example of how the eyes and nervous system can fool our brains. But if the experiment is done correctly the imaginary lines are very clear.

The Austrian physicist, philosopher, and psychologist, Ernst Mach, first reported these bands to scientists a hundred years ago, and formulated a principle for the effect. A technical article touching on Mach bands appeared in *Scientific American,* June 1972.

Try seeing the bands around your shadow on concrete on a sunny day. Sometimes they are more clearly seen if the body is moved.

POLARIZED LIGHT

NEEDED: A pane of window glass, Polaroid sunglasses, a

light source such as sky or clouds.

EXPERIMENT: Hold the glass at about the angle shown in the drawing. Look at the glass through a sunglass lens, and rotate the lens, all the time experimenting with different angles of the window glass. A position will be found where light reflected from the sky or clouds is almost entirely cut off.

REASON: When the glass pane is held at the correct angle to the incoming light it takes out nearly all the light except that vibrating in one direction—that is, it polarizes the light. The Polaroid sunglass lens does the same thing as light is viewed through it.

The light reflected from the pane can be seen through the lens when the lens is in position to allow the polarized light from the pane to pass through it. If the lens is rotated 90 degrees it cuts off the polarized light from the pane.

A good encyclopedia gives lengthy explanations of this phenomenon.

Polarized Light

WATER STARS

NEEDED: A bright shaded light, medicine dropper, water, darkness.

Water Stars

EXPERIMENT: Shine the light down a stairwell, and drop water from the light down (place a towel below to catch the drops). Reflections from the drops will shine like stars.

REASON: Light hitting a drop reflects in all directions from the surface of the drop, and is refracted in all directions as it passes through. The drop might be considered to be a million reflectors and a million lenses.

A few of the ways in which the drop reflects and refracts the light are shown in the diagram of a drop, which need not be perfectly round.

Chapter 12

Household Hints

QUENCH THE FIRE

EXPERIMENT: Sprinkle baking soda on a grease fire to extinguish the flames.

REASON: Baking soda is $NaHCO_3$. When two molecules or any multiple of two are heated, they break up into three fire-killers, as follows: $Na_2CO_3, + H_2O + CO_2$.

In other words, the soda becomes sodium carbonate, a solid which coats the burning grease and helps prevent burning, water which cools the burning grease, and carbon dioxide which helps smother the flames by cutting off the supply of oxygen. This household hint works!

CLEAN COPPER OR BRASS

NEEDED: Household ammonia.

EXPERIMENT: To clean copper or brass, soak in household ammonia, full strength, until the metal is clean. Polish with a cleaning pad of the commercial variety made from very fine steel wool and soap. Do not use regular steel wool; it scratches.

REASON: The corrosion or blackening of the copper or brass is made up of both oxide and dirt or smoke. These do not dissolve readily in water.

The ammonia forms a complex "ion" with the corrosion—a charged group of atoms that will dissolve in water and so may be washed away. The copper ammonia ion is bright blue. Observe the

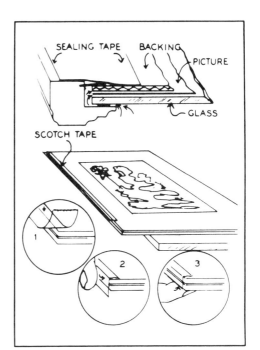

Those Dark Streaks

development of this color while the copper is in the "household ammonia" solution.

THOSE DARK STREAKS

PROBLEM: In the usual picture-framing job, sealing tape or paper is pasted over the back of the frame and the picture backing. Smoke and fine dust can enter between the picture and the glass as shown by the arrows in the upper drawing. The sealing actually does not seal anything.

AN IMPROVEMENT: Place the picture, the glass, and the backing on the edge of a table, and seal them together with cellophane tape as shown in drawings 1, 2, and 3. If care is taken, even the corners can be rather well sealed in this way, so that no dust and smoke can enter except through the pores of the picture backing.

This eliminates much of the dark streaking on pictures that hang for long periods of time on the wall.

PICKLES

NEEDED: Fresh cucumbers, salt water, alum water.
EXPERIMENT: Soak the sliced cucumbers in salt water, and

they will get very limber, because the salt water draws water out of the cells of the pickles by osmosis. This is the process by which a less concentrated solution will flow through a membrane into a more concentrated solution.

Then soak the slices in the alum water. The slices will again become firm.

REASON: The alum solution is a very dilute solution. Water passes into the cells of the cucumber producing a firmness known as "osmotic turgescence."

BUGS AWAY!

EXPERIMENT: When washing greens the tiny bugs wash off much better if salt is added to the water.

REASON: Salt is objectionable and irritating to most insects and other small animals. Most bacteria cannot live in a salt solution, therefore salt can be used to preserve food, by preventing spoilage from bacteria.

Salt solutions coming in contact with cell walls can draw water from the cells by exosmosis, and thus kill the cells. This can be felt in the mouth after eating something very salty. The lips seem to "draw" as their cells lose water.

A WATER RACE

NEEDED: Two pieces of cloth and two small puddles of water.

EXPERIMENT: Wet one of the cloths and squeeze it out. Place both cloths down on the puddles, and it will be seen that the damp cloth soaks up water much faster than the dry one.

REASON: The dry cloth has a considerable air film which prevents the entrance of water. The partly damp cloth has no such air film and entrance to the pores of the cloth is much more rapid as a result.

POT HOLDERS

NEEDED: Two pot holders.

EXPERIMENT: Pick up a hot pan with a dry pot holder and no burn is felt. Pick up the same pan with the wet pot holder and the hand is likely to be burned.

REASON: A dry pot holder consists of fibers with much insulating air space between. Air does not conduct heat very well; water conducts it much better. When the pot holder is wet, water takes the place of the air spaces and will conduct the heat through to

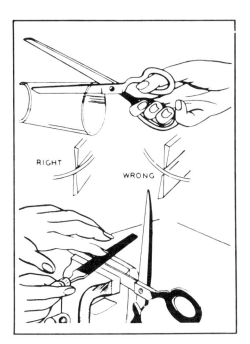

Don't Cut the Glass!

RIGHT WRONG

the hand. It is possible, if the pan is very hot, that enough steam will be produced to go through the wet cloth to burn the hand.

DON'T CUT THE GLASS!

THE SUPERSTITION: An old-wives method for sharpening scissors consists of rubbing them on a glass as if trying to cut it.

CONCLUSION: The scissors may be sharpened somewhat if care is taken that the cutting edges rub against the glass in the correct way shown. If the cutting edges rub the glass incorrectly, the procedure will dull the scissors.

A more reliable way is to place the scissors in a vise and file them as shown in the lower drawing.

HOW TO CLEAN EYEGLASSES

NEEDED: Soap, running water, a freshly laundred cloth.

EXPERIMENT: Wet the glasses, then rub soap on the lenses with the fingers. Rinse the soap off carefully, hold to a light, and wipe them with the cloth.

OBSERVATION: If the lenses are rubbed before washing, oil on them will spread and leave the lenses coated or streaked with it. Also, gritty dirt can scratch the glass if it is not washed off before they are rubbed. This can be shown by putting some sand on a scrap of window glass and rubbing it.

Never lay the glasses down so that the glass touches a tabletop or other hard surface. Gritty dirt is certain to scratch them if this habit is allowed.

Soggy Pancakes

SOGGY PANCAKES

NEEDED: A hot pancake and a cold plate.

EXPERIMENT: Place the hot cake on the cold plate. Lift it after a few seconds; there will be water in the plate, enough to make the hot cake soggy.

REASON: Heat drives water out of the dough in the form of vapor. After the cake is taken from the griddle the heat inside it is still driving out moisture, which condenses on the cold plate.

The best pancake bakers place the cakes on a hot plate. Moisture does not condense on the hot plate, but continues to evaporate into the air. The situation is one of drying the cake rather than wetting it.

LUMPY CUSTARD

NEEDED: The custard ingredients.

Lumpy Custard

EXPERIMENT: When using egg yolks to thicken custard, beat the yolks, then add several tablespoonfuls of the hot custard to the yolks, and mix. This warms the yolks and is called "tempering." It must be done before the yolks are added to the custard to prevent the formation of lumps in the finished custard.

REASON: If egg yolk is heated too quickly, as by placing it in hot custard, it cooks into hard lumps as it would if placed in a hot pan. It is the protein that hardens in this way.

PICKLED EGGS

NEEDED: Hard-boiled eggs, pickling solutin (containing salt and vinegar), a container.

EXPERIMENT: Peel the hard-boiled eggs, and put into the solution. They float. As they soak up the pickling solution they start sinking to the bottom. When on the bottom they are pickled and ready to eat.

REASON: At first the density of the eggs is slightly less than that of the pickling solution, and the eggs float. But the solutions begin to diffuse—the pickling solution goes slowly into the eggs and the egg liquid goes into the solution. The result is that the density of

Pickled Eggs

the eggs increases so it is greater than that of the solution, and the eggs sink.

YOGURT ANYONE?

NEEDED: One quart of milk, three tablespoons of commercial yogurt, a warm oven.

EXPERIMENT: In the top of a double boiler heat the milk to a boiling temperature and then cool it to about 115 degrees F. Add three tablespoons of commercial yogurt, mix well, and pour into jars or custard cups. Place in a pan of water about 115 degrees and keep in a warm oven to keep water at the same temperature until milk sets. You may use the last of this batch to start your next one.

REASON: A special kind of Lactobacillus acidophilus is alive in commercial yogurt. A temperature of 115 degrees Fahrenheit is just right for them to grow and multiply, and it is these little bacteria that make the yogurt. They follow the exponential growth law: one turns to two in a certain time, the two turn to four in a similar time, four turn to eight, etc.

Heating the milk to boiling point kills off any bacteria that might multiply the same way and spoil the special work of the Lactobacillus acidophilus.

Yogurt Anyone?

MAKE TRACING PAPER

NEEDED: Paper, gum turpentine, a brush.

EXPERIMENT: Place a sheet of paper on a piece of window glass or other flat surface, and brush turpentine over it. It becomes translucent, and can be placed over a drawing to be traced. When it dries it resumes its original state, but the traced drawing remains.

REASON: The paper consists of irregular strands of matted cellulose which scatter and reflect light so that dark lines of color or black cannot be seen through an ordinary sheet. The turpentine

Make Tracing Paper

allows a temporary passage of light. Dark lines show dark and white areas are white. But the turpentine evaporates quickly.

Turpentine may be applied with a paper towel, but prolonged contact with the skin should be avoided.

Make Finger Paint

MAKE FINGER PAINT

NEEDED: One package plain gelatin, a half cup of cornstarch, a half cup of mild detergent, food coloring, water.

EXPERIMENT: Mix the gelatin in a fourth of a cup of water, and set it aside. Mix the cornstarch with a three-fourths cup of water, add two cups hot water, and bring to a boil while stirring constantly. Let the mixture boil until it thickens, take away from the heat, and add the gelatin and detergent. Mix well, and pour into different jars where color may be added.

This is a harmless paint for little fingers. Keep it in closed jars for storage. Molds may grow in the gelatin after a time and the paint will be useless.

Chapter 13

Chemistry

A TRICK IN VOLUME

NEEDED: Salt, water, a glass, a dinner plate.

EXPERIMENT: Fill the glass full of water, place it on the dry plate, then pour in a heaping tablespoon of salt. Stir until the salt is dissolved.

Some water will spill out on the plate. After the salt is dissolved, pour the water from the plate back into the glass. It may be possible to pour it all back without spilling any. Or if some spills it will be very little, not nearly as much as the volume of salt dissolved.

REASON: When the salt is dissolved in the water, the molecules of each crystal fit into spaces around each other much as sand would fit in the spaces between marbles in a jar. Much air is between the dry salt crystals but it does not enter the solution. The particles of the salt in solution are about one twenty-five millionths of an inch in diameter.

"BAD BREATH"

NEEDED: Two identical jars, two candles, a soda straw, a match, a pan of water.

EXPERIMENT: Fill a jar with water, insert it in the pan of water, with the neck down, then raise it up so that air from the room may enter and the water may run out. Place the jar on the table, neck down.

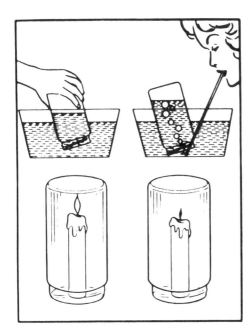

"Bad Breath"

Fill the next jar with water the same way, but blow breath under it with the straw to force out the water. In doing this, take a deep breath, so that the air through the straw will come from the lungs, and not just from the mouth. Place the jar neckdown beside the first.

Light both candles, and place the jars over them at the same time. The candle will burn longer in the jar containing room air.

REASON: The candle flame goes out when the oxygen in the jar is used up. Room air contains about 20 percent oxygen. Exhaled air contains about 16 percent oxygen and 4 percent carbon dioxide. The oxygen not only has been partially replaced with carbon dioxide, but the carbon dioxide interferes with the burning of the candle.

GROW CRYSTALS

NEEDED: Water, Epsom salts, a piece of glass.

EXPERIMENT: Dissolve some of the salts in the water. Pour a few drops of the liquid on the glass, and let it evaporate in a warm place. The salts will be left in the form of beautiful crystals. Examine them under a magnifying glass or a microscope.

OBSERVATION: Exactly why the molecules arrange themselves in definite patterns to form crystals is not known. The shape of the crystal depends upon what substances are used and what the conditions are.

Try other salts, such as table salt or photographer's hypo. Snowflakes are examples of crystals. Note that no two crystals are alike. Some unusual crystal forms may be seen if an egg is broken and its white and yellow allowed to evaporate on a dish.

Crystal formation requires time and room.

A SALT GARDEN

NEEDED: A dish, water, salt, vinegar, some small porous stones or pieces of coal.

EXPERIMENT: Place the stones in the dish. Pour salt into warm water and stir until no more salt will dissolve. Put a spoonful of vinegar in the water, and pour it over the stones in the dish. In a few days, the salt will begin to "grow" and eventually will cover the stones in beautiful crystals.

REASON: The salt water flows up through and over the stones because of capillary action, and as it rises it evaporates leaving the salt behind. The vinegar serves to take away oily spots on the stones that would interfere with the free upward flow of the salt water.

If the solution is left in an open glass the crystals will begin to form above the solution. Crystals there will act as capillaries, allowing more liquid to flow up, more crystals will form, allowing more liquid to flow up. This may continue until there is "growth" of the crystals up and over the edge of the glass.

SEPARATE SALT AND SAND

NEEDED: Salt, sand, water, a cooking vessel, a stove or hot plate.

EXPERIMENT: Mix the salt and sand. Dissolve the salt in hot water. The sand will settle to the bottom of the container, so that the water containing the salt can be poured off. Boil away the water, and the salt will remain.

REASON: The sand does not dissolve in the water as the salt does. The salt does not boil away as the water does.

Will this work with a mixture of sugar and sand? What about sugar and salt?

Separate Salt and Sand

THE SWOLLEN EGG

NEEDED: An egg, vinegar, water.

EXPERIMENT 1: Soak the egg in vinegar until the shell has become soft. Pour out the vinegar, and soak the egg in plain water. In a few days the egg will have become so large that the shell will burst.

EXPERIMENT 2: Without the vinegar treatment, peel just a little square of eggshell off the egg, without breaking the membrane below. By osmosis, the water will go into the egg, pressing upward on the exposed membrane until it breaks.

REASON: The acid in the vinegar combines with the calcium in the eggshell, making the shell soft. When the egg is then immersed in water, the water will go through the shell into the egg until the shell bursts. This is the principle of osmosis, in which a less concentrated solution will go through a membrane into a more concentrated solution more than in the reverse direction.

SALT AND VINEGAR ON COPPER

NEEDED: A copper cent, vinegar, salt, a plate.

EXPERIMENT: Sprinkle a little salt on the coin, pour on some vinegar, and the coin will be cleaned beautifully.

Let it stand for a while, and the copper coin begins to corrode and turn green.

REASON: Here is a little simple chemistry. Salt, or sodium chloride, plus hydrogen acetate in the vinegar, gives us sodium acetate plus hydrogen chloride, or hydrochloric acid.

The hydrochloric acid is strong, and if the bronze cent containing 95 percent copper, 3 percent zinc, and 2 percent tin comes into contact with it and the newly formed salt (sodium acetate), it is cleaned rapidly.

The cleaning process leaves the surface in porous "active" condition so that it quickly corrodes by combining with water, and oxygen and carbon dioxide in the air.

TECHNICOLOR CABBAGE

NEEDED: A cooked red cabbage and various substances for testing.

EXPERIMENT 1: Pour reddish-purple juice from the cooked cabbage into various containers. It will change its color to reveal whether a substance is acid or base.

Lemon juice, which is acid, will turn the solution pink. Baking soda, which is base, will turn it green. Try other substances.

EXPERIMENT 2: Try beet juice as an indicator. Bases will turn it blue.

EXPERIMENT 3: Laundry bleach will take the color out.

REASON: All acids contain hydrogen ions which change the color of litmus and other indicators, including red cabbage juice. Bases contain an oppositely charged negative hydroxyl ion which produces an opposite color effect on litmus and some other vegetable colors including red cabbage juice.

A JAM EXPERIMENT

NEEDED: Blackberry jam or jelly, warm water, substances to test.

EXPERIMENT 1: Do this: Put a spoonful of the jam or jelly into a glass of warm water, dissolve it, and the color is probably red. Put a few drips of ammonia into it, and the color changes to greenish purple.

EXPERIMENT 2: Add just enough ammonia to produce the greenish purple color. Then add lemon juice or vinegar, which are weak acids, and see the color change back to red.

REASON: Many colors from flowers and vegetables can serve as indicators, to tell whether a substance added is acid or alkali. The jelly solution is red when an acid is added to it, greenish purple with an alkali.

The red color of the jam is a natural indicator. An excess of

A Jam Experiment

(OH) ions furnished by an acid gives one color. An excess of (OH) ions furnished by any base gives the other color.

EASY OXYGEN

NEEDED: Hydrogen peroxide (20 percent solution from the pharmacy) and an old dry cell.

EXPERIMENT 1: Put the hydrogen peroxide into a jar (two tablespoonfuls), and sprinkle a tablespoonful of the black mixture from the dry cell over it. Notice the large quantity of bubbles—bubbles of oxygen.

OBSERVATION: The manganese dioxide in the black mixture combines with the hydrogen peroxide to produce manganese hydroxide and free gaseous oxygen, according to the equation $MnO_2 + H_2O_2$ $Mn(OH)_2 + O_2$.

MnO_2 is also a catalyst to speed up the decomposition of the unstable H_2O_2 which decomposes as follows:

$$2H_2O_2 \xrightarrow{MnO_2} 2H_2O + O_2$$

EXPERIMENT 2: Test to see if oxygen is in the jar by placing a glowing splinter in the jar. The oxygen will make the splinter burst into flame.

SPACE PROBLEM

NEEDED: A small glass, a larger one, a wax marking pencil, some water, some rubbing alcohol.

EXPERIMENT: Fill the smaller glass full of water twice and pour the water into the larger glass. Mark carefully on the glass where the surface is. Empty the big glass.

Now fill the smaller glass once with the rubbing alcohol and once with water and pour into the large glass. Be careful each time to fill the smaller glass to the same degree of fullness. The surface will be below the mark, showing that the mixture does not occupy as much space as the water.

REASON: The molecules of water and alcohol are not the same size, and so the smaller molecules can fit somewhat into the spaces between the larger ones. An experiment to demonstrate this: pour equal volumes of marbles and sand into a container, and see how the sand fits into the spaces between the marbles.

Burn Flour

BURN FLOUR

NEEDED: Cheesecloth, flour, a lighted candle.

EXPERIMENT: Make a bag by folding a double thickness of cheesecloth, put flour into it, and shake it over the candle flame. The flour will burn in sudden, sparkling flames.

REASON: Flour does not burn readily unless the particles are so separated that oxygen in the air can reach every one. This condition is met as the fine flour dust comes down to the flame. The finer the flour, the more surface is available for this rapid oxidation.

When the proportion of flour particles to air is just right in a large area, an explosion powerful enough to wreck a flour mill can result. The small explosions in this experiment are harmless, except that fire is always dangerous to some extent. Be careful.

A "POP" BOTTLE

NEEDED: A bottle with a cork, water, vinegar, baking soda, paper, a pan or sink to catch the overflow.

EXPERIMENT: Put water and vinegar into the bottle. Roll a little baking soda in the paper (tissue is best). Drop the paper into the bottle, put the cork on, and soon the cork will pop out.

REASON: When the acid of the vinegar and the baking soda are allowed to mix, they combine chemically to produce carbon dioxide gas. The pressure of the gas as it is formed blows out the cork.

Corks may be hard to get. Try this without the cork; the water and bubbles will overflow. Do not use a screw-top bottle.

A Simple Indicator

A SIMPLE INDICATOR

PROBLEM: Make an Indicator.

NEEDED: Slices of fresh beet, water, a pan, heat, a drinking glass.

EXPERIMENT: Put half a glass of water into the pan. Put in three or four slices of beet and boil for five minutes. Let the red liquid cool, and pour some of it into the glass. Test various substances with it to see whether they are acid or alkaline.

If alkaline substances such as alum water or ammonia are added the color turns to yellowish green or brown. If an acid such as lemon juice is added the reddish color returns.

REASON: Indicators are natural colors from plants which show one color with an excess of H^+ ions from an acid and a different color with an excess of $(OH)^-$ ions from a base. Litmus is the most common indicator in laboratories. The exact mechanism of the color change is not thoroughly understood.

Make Carbon Dioxide

MAKE CARBON DIOXIDE

NEEDED: A gallon jug, a beach ball, a cork stopper, a medicine dropper, water, sugar, molasses, yeast.

EXPERIMENT: Mix a cup of sugar, two tablespoons of molasses, and yeast in 3/4 jug of water. Set in a warm place, attach the ball to it after squeezing out the air. Overnight the ball should fill with carbon dioxide.

REASON: Yeast plants growing in the solution change sugar to alcohol and carbon dioxide gas. The gas fills the ball, as it filled the balloon, with this difference: pressure created in the balloon stopped the fermentation process, and the balloon did not expand. The ball could fill because there was no pressure from tight rubber.

Hold the opening of the ball to the mouth or nose and "taste" some of the gas. It may smell like molasses, and will definitely have a tingle as from a soft drink. Pour some of the gas into a jar, lower a candle into it, and the candle flame goes out.

Inset drawing shows how the cork and medicine dripper are placed to fill the ball.

Carbon Dioxide by Fermentation

CARBON DIOXIDE BY FERMENTATION

NEEDED: Gallon jug, yeast, sugar a balloon.

EXPERIMENT, Dissolve yeast and sugar in a gallon of water in the jug. Fit the rubber balloon over the mouth of the jug, and set the experiment in a warm place where the light is not bright. Bubbles will form in the liquid. They are carbon dioxide, and will have enough pressure to partially inflate the balloon. This takes several days.

REASON: Sugar, either $C_6H_{12}O_6$ or $C_{12}H_{22}O_{11}$ in the presence of zymase, a complex of enzymes, from the yeast plants, gives alcohol and carbon dioxide CO_2. The alcohol dissolves in the water while the gas escapes the water and goes into the balloon.

Light and Chemistry

LIGHT AND CHEMISTRY

NEEDED: Construction paper or newspaper, sunlight.

EXPERIMENT: Place the paper in the sun, and place small objects on it. Leave it for several hours or days. The paper will have changed color where the light hit it. Outlines of the objects may be seen.

REASON: Paper has a natural dye which is very pale or it may be treated with coal tar dyes to give brighter color. The dyes disintegrate and the color of the paper changes slowly when oxygen of the air is activated by bright sunlight. Not all chemical changes resulting from light are well understood.

CHEMICAL ACTION

NEEDED: A candle flame or other fire.

EXPERIMENT: Note that heat is produced as a substance is burned.

REASON: A chemical substance such as candle wax has a certain amount of energy per unit of weight. When a chemical action takes place the products of the action may have more energy or less energy than the original substances.

Chemical Action

A "fuel" is a substance which burns by combining with oxygen of the air to produce products which possess less energy per unit of weight than the original fuel. The difference is liberated as heat, light, and/or motion energy.

When a fuel is burned and the energy is liberated the products are generally worthless and may be dangerous. Automobiles use the energy liberated when gasoline is burned, and the worthless products escape through the tailpipe and pollute the air.

BLEACH FLOWERS

NEEDED: Flowers of different kinds, a jar, a string, a plastic lid for the jar, household ammonia.

EXPERIMENT 1: Tie the flowers and suspend them in the jar by running the string through the lid. Pour a little ammonia into the jar and watch colors disappear.

Bleach Flowers

Our red, pink, and purple flowers turned green. White and yellow flowers were not changed.

REASON: Substances that give the petals their color are present along with chlorophyll which is green. The ammonia gas which rises from the household liquid destroys some of the colors but not others. When colors that hide the green chlorophyll are destroyed the chlorophyll color shows.

EXPERIMENT 2: Try pieces of colored paper and cloth, wetting them before placing them in the ammonia vapor. See how many of their colors are changed.

POLLUTION TO ORDER

NEEDED: Photographer's hypo, lemon juice, an old spoon, a candle for heat.

EXPERIMENT: Put a few crystals of hypo into the spoon with a little lemon juice. Heat it and smell cautiously. The unpleasant choking odor is sulfur dioxide, which is one of the undesirable air pollutants coming from some industries.

REASON: Hypo is sodium thiosulfate, essentially $Na_2S_2O_3$ and when heated with lemon juice some of the sulfur (S) and oxygen

(O) atoms unite as the hypo molecules are destroyed by the heat. The sulfur dioxide (SO_2) formed always has a stinking, unpleasant odor, and is poisonous, but the amount produced in this experiment is not dangerous.

Pollution to Order

A BIG BAD ODOR

NEEDED: Photographer's hypo (sodium thiosulfate, an old spoon, a candle for heat.

EXPERIMENT: Put a few hypo crystals in the spoon and hold it over heat until the hypo melts. Continue to heat it, and the odor of rotten eggs will be given off. The odor is that of hydrogen sulfide, and this is a simple chemical experiment.

REASON: There is some water of crystallization in the seemingly dry hypo crystals. When the crystals are heated hydrogen (H) atoms from the water and sulfur (S) atoms from the hypo unite to form hydrogen sulfied (H_2S) molecules.

Rotten eggs also produce H_2S molecules, which always smell the same regardless of how they are produced.

CASE OF THE DISAPPEARING CRYSTALS

NEEDED: Salt or sugar and water.

EXPERIMENT: Dissolve a spoonful of the crystals in a glass of water. They disappear; the water becomes clear. When the water

174

Case of the
Disappearing Crystals

evaporates, the crystals again appear. Where do they go when they dissolve?

REASON: Crystals consist of thousands to millions or more of molecules of a definite chemical composition held together in a definite geometric pattern. When they dissolve they separate into smaller or individual invisible particles distributed in the solution.

When the solution is evaporated the smaller particles again unite in a definite geometric pattern if enough time is allowed, and they become large enough to be seen again.

A FIRE HAZARD

NEEDED: A lighted cigarette and dry grass or pulverized dry leaves.

EXPERIMENT 1: Try to set fire to the dry material by dropping the cigarette into it. It is difficult.

EXPERIMENT 2: Try lighting shavings from a pencil sharpener with a cigarette.

EXPERIMENT 3: Place a lighted cigarette and a lighted cigar on an ash tray. The cigarette will continue to burn; the cigar will not. Cigarettes actually are incendiary bombs. They can set fire to grass and leaves in the open when there is a breeze.

A Fire Hazard

OXYGEN

NEEDED: Three candles, drinking glass, large jar.

EXPERIMENT 1: Light all the candles. Place the jar and glass over two of them. The flame will go out under the glass, and later will be extinguished under the jar.

Oxygen

EXPERIMENT 2: Try keeping insects or small animals in the jar after the flame has gone out.

REASON: The flame must have oxygen in order to burn. As the oxygen is depleted the flame goes out. The flame burns longer under the large jar because it contains more oxygen. It continues to burn when no jar or glass is placed over it because there is plenty of oxygen in the air.

One statement sometimes made in this connection is erroneous: the flame consumes all the oxygen. This is not true. It consumes a large part of the oxygen; after the flame goes out there is enough left under the glass to keep small animals or insects alive for quite a while.

Some fresh air is sucked into the space, too, as the air inside cools and contracts.

Oxidize Iron

OXIDIZE IRON

NEEDED: Two tall jars with lids, a candle, steel wool.

EXPERIMENT: Wet a wad of steel wool and put it into one jar. Close the lids of both jars and let them stand two days. Light the candle, and let it down into the jar without the steel wool. It will burn a few minutes. Light it again, lower it into the jar with the steel wool, and its flame goes out quickly.

REASON: The steel wool rusts, and rusting is slow oxidation. The process uses up oxygen, so that after two days nearly all the oxygen in the jar is used up. There should not be enough to support the burning of the candle wick.

Perform this experiment where there is no breeze to blow the gases of oxidation out of the jar. Some steel wool is covered with an oil film to protect it from rusting in air on the dealer's shelf. To remove the oil, wash the steel wool in varsol then let it dry overnight to get the varsol off.

Make Ink

MAKE INK

NEEDED: Fine steel wool, varsol or kerosene for cleaning it, white vinegar, tea bags, mucilage, containers.

EXPERIMENT: Clean a wad of steel wool with varsol and let it dry overnight. Put the steel wool into a jar, and cover it with vinegar. Set it in a pan of water, hot but not boiling. Put four tea bags into half a cup of water, and boil. Let both solutions cool, mix them in equal amounts, and stir. Dip a finger into the mixture, and mark a large X on a newspaper page. It will gradually show up black.

REASON: Chemical action of the vinegar and iron produces hydrogen and iron acetate. The tea yields tannin. Mixed, they produce ferrous tannate, which is almost colorless but which changes to ferric tannate when dry and exposed to the air. In from three hours to a day the color change in the X should be complete—it should be black. A little mucilage mixed with the newly-produced ink will allow it to flow from a pen as regular ink.

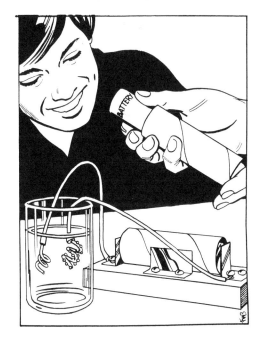

Electrolytes

ELECTROLYTES

NEEDED: A battery of three to six volts, salt, sugar, water, wire, glass containers.

EXPERIMENT: Connect the wires to the battery, bare the other ends of them, and hold the bare ends in water. Nothing happens. Put sugar in the water; nothing happens. Get fresh water, put salt in it, place the bare wires in it, and bubbles form on one wire.

REASON: Salt is a compound of sodium and chlorine. When the salt dissolves some sodium atoms give up electrons to become positive in charge; some chlorine atoms have grabbed those extra electrons and have become negative in charge. These charged atoms are called "ions."

The positively charged ions move to the negatively charged wire where they pick up electrons enough to make them neutral. But sodium metal reacts very rapidly and vigorously with water to give sodium hydroxide, Na(OH), and hydrogen gas. It is the hydrogen gas that is seen.

Pure water and sugar water do not form ions. Those that do and can carry electricity are called "electrolytes." Acids, bases, and salts form ions in many solutions.

Chapter 14

Mechanics

THE RUBBER ERASER

NEEDED: A pencil eraser, a rubber balloon, pencil, paper.

EXPERIMENT: Notice how the pencil eraser will erase the marks, while the rubber balloon has more of a tendency to smear them.

REASON: The pencil eraser has tiny pieces of hardened gritty rubber held in position by the rubber film. As the eraser is moved, the small gritty pieces remove the graphite marks by removing the upper surface of the felted paper. The balloon has only a smooth rubber film which spreads the marks over the surface of the paper.

A JET-PROPELLED BOAT

NEEDED: A piece of wood, a medicine dropper, a balloon, a place to float it.

EXPERIMENT 1: Make the boat as shown. It will be propelled through the water as the air leaves the balloon through the dropper.

REASON: When the balloon is blown up, the stretched rubber pushes on the air inside it. When there is an opening, as through the dropper, the air is pushed out by the balloon and the balloon is pushed by the air causing the boat attached to the balloon to move.

Newton's Third Law: "Every action has an equal and opposite reaction."

EXPERIMENT 2: Sail the boat so that the end of the dropper

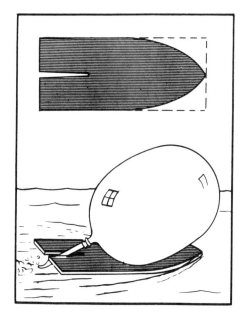

A Jet-Propelled Boat

is below the water surface, then again when the opening is above. The boat is likely to go farther when the opening is below the water, not because it pushes against the water more than against the air, but because the air leaves the balloon faster when allowed to exhaust above the water. At the faster speed, the wind and water resistance is greater for the boat.

MUSCLES AND LEVERAGE

NEEDED: A broom and three people.

EXPERIMENT: Have two people hold the broom tightly as shown, so that the end is a foot or more above the floor. Ask them to bring down the broom handle so that it will come quickly into the circle or ring on the floor. You can push the broom aside with one finger so that it will not hit the ring.

REASON: Because of the leverage, a small force exerted by the finger at the end of the broomstick can easily change the direction of motion of the stick. Coordination of the two men has something to do with it also. The two people trying to bring the stick into the ring cannot work together; actually more often, they will find they are pushing against each other.

The greater length of the stick to the child's hands allows him to produce greater torque than the men, who have a length of stick much closer to the center of rotation.

Muscles and Leverage

THE CANDLE SEE-SAW

NEEDED: A candle, a long needle, matches, a newspaper to catch the wax.

EXPERIMENT: Push the needle through the candle, then light it at both ends. It will "see-saw."

REASON: The wax will melt faster on the lower end because the flame from the wick will heat it more. As that end gets lighter,

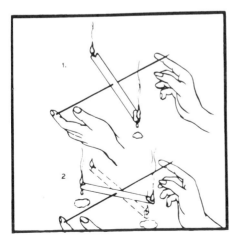

The Candle See-Saw

the other end will come down and start to melt faster. The process is repeated, making the candle see-saw slowly.

YOU AND A HORSE

NEEDED: A ruler, a scratch pad, a pencil.

EXPERIMENT: Measure the height of the stairs or steps, then find out how much energy is used in climbing them.

METHOD: Multiply the height by your weight to get foot-pounds of energy. Suppose you weigh 110 pounds and the stair is ten feet high, you have used 1110 foot-pounds of energy. To change this into horsepower, another figure must be added: time. Horsepower is 33,000 foot-pounds per minute, or 550 foot-pounds per second. If you climb the stairs in five seconds, then your power is 1100 divided by five times 550, or four-tenths of a horsepower.

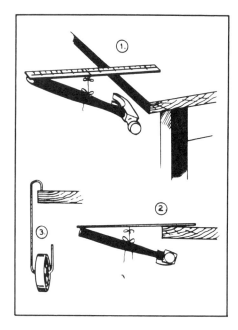

Defying Gravity

DEFYING GRAVITY

NEEDED: A hammer, a ruler, a string, a table edge.

EXPERIMENT: Tie the ruler and hammer together as shown, and they will hang from the table in what will look like a most precarious manner.

REASON: Most of the weight is in the hammer head, and if the ruler is moved along the table edge until the center of weight of the assembly is directly under the edge, the balance point will be easy to find. It is as if the weight were hanging straight down from the table. The center of gravity of the assembly must be on the table side of the table edge.

Multiplied Muscle Power

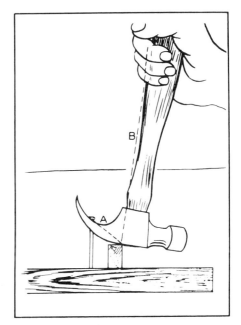

MULTIPLIED MUSCLE POWER

NEEDED: A hammer and a nail, a piece of wood, a small block of wood.

EXPERIMENT: Try pulling the nail with the nose of the hammer against the wood. Then place the small block under the hammer as shown, and the nail will be pulled easily.

REASON: Note the two broken lines in the drawing. If the distance of line A is one inch, and the distance of line B is 10 inches, the pull on the handle places about ten times as much pull on the nail. If 40 pounds of pull is exerted on the handle, about 400 pounds of pull is exerted on the nail. The pulling force applied to the handle will move ten times as far as the nail moves.

The hammer is a form of lever.

WEIGHT LIFTING

NEEDED: A weight.
EXPERIMENT: Lift the weight as in the drawing at left, and

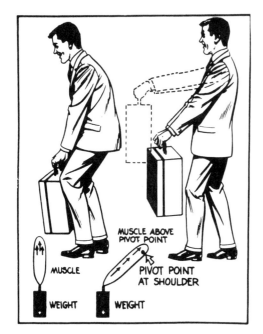

Weight Lifting

MUSCLE ABOVE PIVOT POINT

MUSCLE

PIVOT POINT AT SHOULDER

WEIGHT

WEIGHT

it is easy. Try to lift it as shown in the drawing at right, and it is difficult or impossible.

REASON: In the left drawing, the muscle tension (force upward) and the weight act along the same line and are equal. Both are fairly small.

In the right drawing, where the arm is extended, the muscle tension times its distance from the pivot point must equal the weight times its distance from the pivot point (length of arm). Since the weight is far from the pivot point and the muscle close to the pivot point, the muscle tension must be many times the weight, if it is to support the weight.

The lower drawings show this in an over-simplified manner. The pivot point is the shoulder joint.

GLAMORIZING THE WEDGE

DICTIONARY DEFINITION: A wedge is a piece of wood or metal, small at one end and larger at the other, used for rending or compressing.

COMMENT: A wedge is a type of inclined plane which is pushed into an object to cut or split it. The smaller the angle of the

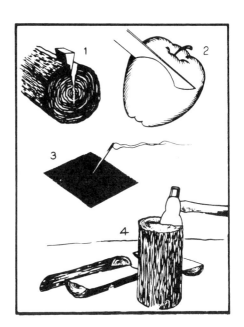

Glamorizing the Wedge

wedge, the easier it is to cut the object; therefore a sharp knife cuts better than a dull one.

The push required to move a wedge into an object is not easy to determine because of friction.

The wedge is used by carpenters and woodsmen in the form of the ax, chisel, plane, and nail. The farmer turns his soil with a wedge—the plow. A rotating wedge or cam is used to push up the valve rods in automobile engines. A needle is a wedge, too.

TRANSVERSE WAVE MOTION

NEEDED: Hose or rope, a large pan of water, a cork, small stone.

EXPERIMENT 1: Move the hose or rope up and down to make transverse waves as shown in the drawing.

EXPERIMENT 2: Drop the stone into the water. The floating cork does not move outward with the waves, but moves up and down.

REASON: The wave form moves down the length of the hose as the hose itself moves only up and down.

Similarly, the wave form from the disturbance in the water moves over the surface of the water as the water particles move up and down. The cork shows that this is true, as it moves up and down.

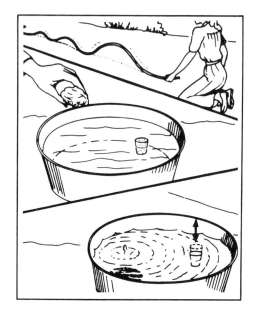

Transverse Wave Motion

THE SQUEEZE BOTTLE

NEEDED: A flat-sided bottle filled with water and a thin card or paper to cover the bottle mouth.

EXPERIMENT: Place the paper over the bottle mouth, invert the bottle, and the water will not run out because atmospheric pressure holds it in. This is shown in A. Squeeze the bottle on the flat sides and some of the water will come out.

REASON: The bottle, while made of glass, actually is elastic,

The Squeeze Bottle

and may be distorted by squeezing so that some of the water is forced out. If it is held long enough for warmth from the hands to go through the glass, the expansion of the water due to the heat will force a little more out.

BUOYANCY

NEEDED: A rubber balloon and a pan of water.

EXPERIMENT: Blow up the balloon, and see how difficult it is to push it under the water.

REASON: The water in the pan is much heavier than the balloon and the air in the balloon. The force required to move the balloon down into the water is equal to the weight of the water raised or "displaced" by the balloon. Archimedes discovered this principle more than 2200 years ago.

If we wanted to be strictly accurate, we would have to say "the additional force required," because the weight of the balloon, although slight, would displace a small amount of water.

ELASTICITY

NEEDED: A rubber balloon.

EXPERIMENT: Inflate the balloon with the breath. It is difficult or impossible with some new balloons. Now stretch the

Elasticity

balloon several times with the fingers. It is easier to inflate it with the breath now.

REASON: Rubber, while usually thought of as very elastic, is not one of the most elastic substances. When the balloon is stretched with the hands it never goes back to the tough, tight, difficult state it was in before stretching.

A definition of "elastic" is: "Able to return immediately to its original size or shape, after being altered by sqeezing, stretching, compressing, or bending." Rubber does not fit that definition. Rubber is very "stretchable," however, and this is one of the characteristics of elasticity.

More Elasticity

MORE ELASTICITY

NEEDED: Rubber band, strips of scrap glass, a vise with plastic jaw inserts, gloves, protective goggles.

EXPERIMENT: Clamp the glass in the vise rather tightly. The plastic will help prevent it from cracking. Wiggle one of the glass strips back and forth. It always returns to its original shape.

Now measure the length of the rubber band, and stretch it as far as possible several times. Measure it again. It is longer than before.

A definition of "elastic" is: "Able to return immediately to its original size or shape, after being altered by squeezing, stretching, compressing, bending, etc." Glass is more elastic than rubber.

Get two long strips of scrap glass from a hardware store. Place on a book or two and press down on the center of one. It will bend a surprising amount before breaking.

Friction of Water

FRICTION OF WATER

NEEDED: Two similar jars with tight lids, one half filled with water, and a slightly inclined board.

EXPERIMENT 1: Release the jars at the same time at the top of the incline. The one containing water will roll faster at first, but the other should roll farther when it reaches the level of the floor.

EXPERIMENT 2: Try this by filling one jar half full of water and filling the other completely with water. Which will roll farther?

EXPERIMENT 3: Fill one jar completely with water; leave the other without water. See which will roll farther.

REASON: Friction between the water and the sides of the glass jar make the jar slow down. Air in the "empty" jar produces no such friction.

Rolling Right Along

ROLLING RIGHT ALONG

NEEDED: Two cans or jars with lids and sand.

EXPERIMENT 1: Fill the jars with sand. Stand one on end and pull it along on the table or floor. Roll the other. Note that the can on end is much harder to move along.

EXPERIMENT 2: Pour out half the sand in one jar, and try to roll it. It will not roll smoothly at all.

REASON: When the can of sand is pulled along, dragging on the table, small imperfections in the touching surfaces are always catching on one another. The friction produced must be overcome before the can can move. When the can is rolled it travels over the bumps rather than through them.

When a can is half empty, rolling the can part way builds up the sand on one side. If released the can will roll backward until the sand is balanced on both sides of the can. Continuous rolling involves continual lifting of part of the sand. This friction is between the sand and the side of the can, and between sand and sand.

THE BRAKE

NEEDED: Cardboard, a pencil, something for marking a circle.

EXPERIMENT: Cut a circle from the card and push the pencil

The Brake

through the center. Twirl the newly made "top" on the table. It will turn for several seconds. But if the pencil is touched between finger and thumb it stops quickly.

REASON: The friction between the pencil and finger slows the top down. If not touched the friction between the pencil and the table and between the card and the air will gradually stop the motion.

The fingers illustrate the principle of the brake in the automobile. When the brake pedal is pushed a material is pushed against a turning wheel or drum, causing friction which tends to slow the motion.

THE WHEEL

NEEDED: Meter stick or yardstick, toy wagon with rubber tires, a flashlight battery, foot rule.

EXPERIMENT: Place a marker on the ground exactly under the rear axle of the wagon. Place the end of the meter stick exactly over the axle, on the tire. Hold the stick on the rubber as the wagon moves along, letting the stick move forward without slipping as the wheel turns. Stop when the other end is exactly over the axle.

The Wheel

Measure the distance from the marker to the axle. Guess: will it be the length of the stick? Half the length of the stick? Or twice the length of the stick?

REASON: The top of the wheel moves twice as fast as the axle when the wagon is pulled along. The point of the tire touching the ground does not move at all—at the instant it is in contact with the ground.

This may be tried on a tabletop with a rule and round object. The pencil shown is the marker.

THE BRIDGE

NEEDED: Strips of cardboard about 4 × 12 inches, books, jars, sand.

EXPERIMENT: Make a bridge by putting two strips of card between stacks of books as shown in the upper drawing. It will not hold up an empty jar. Make the bridge as shown in the lower drawing, place the jar on it, pour sand into the jar. The bridge is stronger.

REASON: Materials may be stressed several ways: compression or pushing together, tension or pulling apart, flexure which

194

The Bridge

means bending, torsion or twisting, shearing or slicing apart.

The cards will bend easily when weight is put on them as in the first try. But when one card is placed below, it is pushed on with a compressionable force. It is quite strong when compressional or tensile forces are applied, and so holds up the upper card which rests on it.

Index

OTHER POPULAR TAB BOOKS OF INTEREST

TAB   TAB BOOKS Inc.

Blue Ridge Summit, Pa. 17214

Send for FREE TAB Catalog describing over 750 current titles in print.